The World Bank
INTERNATIONAL BANK FOR RECONSTRUCTION AND DEVELOPMENT
INTERNATIONAL DEVELOPMENT ASSOCIATION

1818 H Street, N.W.
Washington, D.C. 20433
U.S.A.

(202) 477-1234
Cable Address: INTBAFRAD
Cable Address: INDEVAS

April 15, 1985

Dear Madam/Sir:

Subject: UNDP/World Bank Integrated Resource Recovery Project (Waste Recycling--GLO/80/004, GLO/84/007)

In 1981, a three-year Global Research and Development Project on Integrated Resource Recovery (Waste Recycling) was initiated as Project GLO/80/004 by the United Nations Development Programme through its Division for Global and Interregional Projects. The World Bank, through its Water Supply and Urban Development Department (WUD), agreed to act as executing agency.

The primary project goal is to achieve economic and social benefits through sustainable resource recovery activities in the developing countries by the recycling and reuse of solid and liquid wastes from municipal and commercial sources.

Increasing recognition of the need for technical and economic efficiency in the allocation and utilization of resources and the role that appropriate recycling can play in the water and sanitation sector has led to the inclusion of this project in the formal activities of the United Nations International Drinking Water Supply and Sanitation Decade.

The recycling of human wastes to add nutrients to and improve the protein production in aquaculture ponds is an ancient practice. In its modern form, the reuse of wastewater effluents for aquaculture, followed by irrigation of crops, offers attractive benefits, including the increase in water supplies for productive agricultural use and the addition of valuable fertilizers and micronutrients to maintain aquaculture growth and later soil fertility, while contributing to the reduction of pollution of surface water sources.

Possible negative effects to the public consuming the fish raised in aquaculture ponds is being studied in parallel at our project site in Lima, Peru; findings will be reported at a later date.

This paper presents issues of waste-fed aquaculture--sanitary, commercial, and public health aspects, particularly for conditions relevant to developing countries. This study has been carried out for the World Bank/UNDP by Dr. P. Edwards, from the Asian Institute of Technology (AIT), Thailand.

Comments and remarks on this report are most welcome.

S. Arlosoroff

Saul Arlosoroff, Chief
Applied Research & Technology
(UNDP Projects Management)
Water Supply & Urban Development Department

Enclosure

INTEGRATED RESOURCE RECOVERY PROJECT

UNDP/World Bank GLO/80/004

Recycling from Municipal Refuse (Technical Paper No. 30), 1984. S. Cointreau et al.

Remanufacturing (Technical Paper No. 31), 1984. R.T. Lund, MIT.

Municipal Waste Processing in Europe (Technical Paper No. 37), 1985. J.G. Abert.

FORTHCOMING PUBLICATIONS:

Health Effects of Wastewater Irrigation and Their Control in Developing Countries, 1985. H. Shuval et al.

RURAL WATER SUPPLY HANDPUMPS PROJECT

UNDP/World Bank INT/81/026

Rural Water Supply Handpumps Project Report No. 3: Laboratory Testing of Handpumps for Developing Countries, Final Technical Report, 1984. Consumers' Association Testing and Research Laboratories.

Rural Water Supply Handpumps Project Report No. 4: Progress Report on Field and Laboratory Testing (Technical Paper No. 29), 1984. S. Arlosoroff et al.

TO BE PUBLISHED IN JUNE 1985:

Rural Water Supply Handpumps Project Report No. 5: Proceedings of Workshop
Ministry of Machine Building Industry, United Nations Development Programme, World Bank Workshop on the State of the Art and Application of Handpumps and Water Supply and Sanitation.
Changsha, Peoples Republic of China
August 15-21, 1984.

INFORMATION AND TRAINING FOR LOW COST WATER SUPPLY AND SANITATION PROJECT

UNDP/World Bank INT/82/002

Water Supply and Sanitation Project Preparation Handbook, 1983.

Volume 1: Guidelines (Technical Paper No. 12). Brian Grover.

Volume 2: Case Studies--Identification Report for Port City, Immediate Improvement Project for Port City, Pre-Feasibility Report for Farmville, Pre-Feasibility Report for Port City (Technical Paper No. 13). Brian Grover, Nicholas Burnett, Michael McGarry.

Volume 3: Case Study--Feasibility Report for Port City (Technical Paper No. 14). Brian Grover, Nicholas Burnett, Michael McGarry.

Low Cost Sanitation Publications (TAG)

Ten working papers and technical notes on different aspects of low cost sanitation, latrines construction and others.

Integrated Resource Recovery

UNDP Project Management Report Number 3

INTEGRATED RESOURCE RECOVERY SERIES
GLO/80/004
Number 3

This is the third in a series of reports being prepared by the Integrated Resource Recovery Project as part of a global effort to realize the goal of the United Nations International Drinking Water Supply and Sanitation Decade, which is to extend domestic and community water supply and sanitation services throughout the developing world during 1981 to 1990. The project objective is to encourage resource recovery as a means of offsetting some of the costs of community sanitation.

Volumes published to date include:

RECYCLING FROM MUNICIPAL REFUSE: A State-of-the-Art Review and Annotated Bibliography

REMANUFACTURING: The Experience of the United States and Implications for Developing Countries

Other proposed volumes include reports on:

- Anaerobic Digestion
- Composting
- Demand Analysis
- Economic Analysis
- Effluent Irrigation
- Transferable Technologies
- Ultimate (marine) Disposal

and a series of case studies of various projects throughout the world.

Series cover design (clockwise from top): Aquaculture using wastewater yields about 8 tons of fish per hectare per year in India. Biogas is produced from organic wastes in India. Sullage from a shower is used to irrigate a garden in the Sudan. The original value added to aluminum is captured by using waste oil to melt scrap and then pouring new ingots in Egypt. A "state-of-the-art" plant, built to demonstrate the pyrolysis of garbage to make fuel oil, has been shut down temporarily because of excessive operation and maintenance costs in the United States. Paper is recycled in a factory of the Shanghai Resource Recovery and Utilization Company in China.

WORLD BANK TECHNICAL PAPER NUMBER 36

Aquaculture

A Component of Low Cost Sanitation Technology

Peter Edwards

The World Bank
Washington, D.C., U.S.A.

Manufactured in the United States of America
First printing April 1985

This is a document published informally by the World Bank. In order that the information contained in it can be presented with the least possible delay, the typescript has not been prepared in accordance with the procedures appropriate to formal printed texts, and the World Bank accepts no responsibility for errors. The publication is supplied at a token charge to defray part of the cost of manufacture and distribution.

The World Bank does not accept responsibility for the views expressed herein, which are those of the author(s) and should not be attributed to the World Bank or to its affiliated organizations. The findings, interpretations, and conclusions are the results of research supported by the Bank; they do not necessarily represent official policy of the Bank. The designations employed, the presentation of material, and any maps used in this document are solely for the convenience of the reader and do not imply the expression of any opinion whatsoever on the part of the World Bank or its affiliates concerning the legal status of any country, territory, city, area, or of its authorities, or concerning the delimitation of its boundaries or national affiliation.

The most recent World Bank publications are described in the annual spring and fall lists; the continuing research program is described in the annual *Abstracts of Current Studies.* The latest edition of each is available free of charge from the Publications Sales Unit, Department T, The World Bank, 1818 H Street, N.W., Washington, D.C. 20433, U.S.A., or from the European Office of the Bank, 66 avenue d'Iéna, 75116 Paris, France.

Peter Edwards is an associate professor of aquaculture at the Asian Institute of Technology in Bangkok and a consultant to the World Bank.

Library of Congress Cataloging in Publication Data

Edwards, Peter, 1926-
Aquaculture : a component of low cost sanitation technology.

(World Bank technical paper, ISSN 0253-7494 ; no. 36)
(Integrated resource recovery series ; no. 3)
Bibliography: p.
1. Recycling (Waste, etc.) 2. Sanitation engineering. 3. Aquaculture--Waste disposal. I. Title. II. Series. III. Series: Integrated resource recovery series ; no. 3.
TD794.5.E37 1985 628.3'6 85-6555
ISBN 0-8213-0527-1

ABSTRACT

This paper discusses all phases of aquaculture, including commercial viability, sanitary and biological considerations, public health, financial/economic and sociological aspects. Current studies are detailed and options are discussed for their potential applicability to developing countries, considering requirements for capital and labor skills as well as physical needs such as land.

ACKNOWLEDGMENTS

The writer is deeply indebted to Dr. Michael McGarry, who in his former capacity as Programme Officer with the International Development Research Council, Canada, first made available research funds to study aquaculture excreta reuse systems at the Asian Institute of Technology, Bangkok. Thanks are also given to the IDRC for providing the research funds to the writer for more than four years. Dr. Chongrak Polprasert, who works with the writer in waste recycling at AIT, is thanked for valuable discussions concerning the engineering aspects of excreta reuse. Finally, thanks are due to the Overseas Development Administration, London, which seconds the writer to the AIT, and which is currently sponsoring a research programme at AIT in aquaculture excreta reuse.

CONTENTS

LIST OF FIGURES

FOREWORD

This report is one of a series being published by the World Bank as executing agency for the UNDP Integrated Resource Recovery Project (GLO/80/004), aiming at improved sanitation through the integration of recycling and resource recovery methods.

The use of aquaculture in an integrated resource recovery program which includes improved sanitation as a goal has been demonstrated to be sustainable using human and other resources. Waste fed aquaculture is the use of human excreta or wastewater as fertilizer or nutrients for fish or water plants. Systems have been developed using raw sewage directly, or in combination with other treatment methods. Because of the potential for economic return, which can reduce the overall capital requirements for sewage treatment, the methods discussed should be of value to decision makers in developing countries and other interested agencies.

The use of excreta and wastewater to provide fertilizer, nutrients for fish, or a gaseous fuel can give a positive economic incentive for improving sanitation in developing countries. The economic benefits from using excreta in these ways are often more tangible than the benefits to public health, and may therefore provide stronger motivation for better sanitation.

We look forward to receiving any comments and case study information, from which future editions will benefit. Please send them to Mr. S. Arlosoroff, Chief (WUDAT), World Bank, Washington, D.C. 20433, U.S.A.

Fish harvest in an experimental fish pond fed with raw sewage at Rahara, West Bengal, India.

CHAPTER I. INTRODUCTION

The decade 1981-1990 has been declared the International Drinking Water Supply and Sanitation Decade by the United Nations. One of the two major goals is to provide sanitary disposal of human wastes for the population of the developing countries, a difficult task since today 1.1 billion* people in developing countries lack adequate sanitation. Preventable waterborne diseases alone or in conjunction with malnutrition cause the deaths of 25,000 people/day, a total of approximately 9 million people per year.

Multi-yearly efforts of the World Bank have led to a set of guidelines and reports for the implementation of low cost water supply and sanitation technologies for thóse who cannot afford the expensive and complicated conventional waterborne sewerage techniques. The identification of appropriate sanitary technologies that are lower in cost than sewerage still does not solve the major problem that sanitation in many cases is not financially remunerative. Low income householders in developing countries may not only be unwilling, but also unable to pay for adequate sanitation. Central and municipal governments may be morally committed to improve sanitation, but they may not be financially able to subsidize the systems indefinitely. Human excreta should be regarded as natural resources to be conserved and reused rather than discarded, and thus provide positive economic incentives for the improvement of sanitation. Economic benefits of excreta reuse are more tangible than benefits to public health and may therefore provide a stronger motivation for further investments in sanitation. In addition, by channelling human wastes through an organized collection and reuse system, the level of potential contamination and exposure would be restricted, and should be beneficial in terms of health. The introduction of any treatment processes in the recycling process would further reduce health risks in any form of disposal or re-use.

Millions of people in Asia depend on waste recycling for the treatment of excreta, and the provision of fuel, soil nutrients and fish proteins. Traditional waste recycling technologies should be studied, and improved if necessary to safeguard public health, but the principle of resource recovery should be inviolate. Furthermore, it is often not appreciated that there is a considerable amount of human waste recycling in the West: the widespread use of sewage sludge in agriculture in Europe and the USA, and the use of sewage in fish culture in certain parts of Europe, and other regions.

The objective of this report is to outline the feasibility of integrating aquaculture into low cost sanitation systems. Examples of existing, commercially viable, human waste reuse systems involving aquaculture are described. The feasibility of linking aquaculture to various low cost sanitation technology options is discussed with an outline of the major biological considerations in waste fed aquaculture systems. Public health, sociological, and economic aspects of aquatic waste reuse schemes are considered.

*billion = thousand million

This report will be supplemented by a report on a waste fed aquaculture research and demonstration project, undertaken within the UNDP/World Bank Resource Recovery program in Lima, Peru, and other studies in waste fed aquaculture.

CHAPTER II. COMMERCIALLY VIABLE AQUACULTURE EXCRETA REUSE SYSTEMS

The use of excreta in aquaculture is an age-old practice, particularly in Asia, but fish culture using water enriched with human excreta was widely practiced by European monasteries in the Middle Ages. However, most of the excreta reuse systems today are located in Asia, although there are commercially viable systems in operation in Europe, particularly in the Federal Republic of Germany.

To assess the potential for excreta reuse in aquaculture, the first step should be to study those systems which are in actual operation in the world today. The purpose of this chapter is to describe existing excreta based aquaculture systems and thus to demonstrate that the concept of excreta recycling is a viable commercial proposition in diverse societies in both developing and developed countries.

DRY SYSTEMS

These systems utilize nightsoil or fecally contaminated surface water for aquaculture and are differentiated from "wet" sanitation because of the relatively little amount of water used.

China

The Chinese utilize excreta more fully than any other culture. Nightsoil is collected in both rural and urban areas and used largely in agriculture, although significant amounts are used for fish culture. The nightsoil may be transported by carts, vacuum trucks, or by boats (Fig. 1). Overhanging latrines are a common sight on fish ponds in Guangdong Province in southern China. From 1952 to 1966, between 28 and 38% of the nutrients applied in agriculture came from nightsoil. In 1952, 176 million tons wet weight of nightsoil, about 70% of the total excreta in China, were recycled and this had increased to 299 million tons wet weight or 90% of the total by 1966 (McGarry, 1976). Without excreta recycling, the Chinese might not be able to maintain their agricultural production.

There are few specific data available on recycling nightsoil in aquaculture from Mainland China. However, more data are available from Taiwan, China, where nightsoil, and more recently septic tank sludge, are used for both brackish water and freshwater fish ponds. In Tainan, Taiwan, there are several thousand hectares of ponds, about 80% being brackish water ponds. In brackish water ponds, nightsoil is spread on the pond bottom when the ponds are empty during the winter, but in freshwater ponds nightsoil is added at intervals, about 4 to 6 times during the growing season. Supplementary feed such as agricultural by-products and grain are pond inputs besides nightsoil in both Taiwan and on the Mainland.

In Taiwan there is also a different type of excreta recycling system, in which fecally polluted surface water is used to fertilize fish ponds. During the winter when the ponds are drained, polluted water is pumped into a depth of about 30 cm to induce the growth of bottoms (benthic) biomass. When the water has lost

Fig. 1 Ferro-cement boat for the transportation of manure, including nightsoil, to a fish farm in Wuxi, Jiangsu, China.

Fig. 2 Overhanging latrine on fish pond, Pathumthani, Thailand.

its turbidity, the clear water is drained or pumped out and a new supply of nutrient rich polluted water is again pumped in. This operation is repeated 3 or 4 times to ensure the development of a rich growth of benthic organisms on the pond bottom. The water level is increased in March when fish are stocked and polluted water is pumped into the pond once every 3 days to maintain the fertility of the water.

Malaysia and Thailand

The migration of Chinese throughout Southeast Asia has led to the introduction of excreta reuse to these countries, but it occurs on a very reduced scale compared to China. Overhanging latrines are occasionally seen on fish ponds in Thailand (Fig. 2) and Malaysia, and there are a few fish farms in Malaysia that recycle nightsoil collected manually from households and transported to the ponds.

Viet Nam

In Viet Nam various excreta reuse strategies are employed in aquaculture, in addition to overhanging latrines. Human and animal excreta plus vegetable matter may be anaerobically composted in a concrete tank and both the effluent and the composted material added to fish ponds. Pour flush toilets with vaults may be used and both the effluent and vault contents added to fish ponds. The compost from composting toilets, and the urine which is separated from the feces, may both be added to fish ponds.

Bangladesh

There is some use of nightsoil for fish culture. In certain areas, overhanging latrines are constructed in ditches located behind the house, which are practically empty during the dry season. Flood waters fill up the ditch during the rainy season, and the fish that enter with the water grow rapidly in the nutrient rich water, which is continuously enriched from the overhanging latrine. The fish are harvested when the water level falls, following the end of the monsoon season. However, the use of excreta in fish culture in Bangladesh is incidental rather than intentional.

Indonesia

Excreta reuse in fish culture is a traditional part of aquaculture in Indonesia, particularly in West Java. The most common system in use is the overhanging latrine; in West Java about 25% of the fish ponds have latrines and in some districts the number may exceed 90%. The ponds are small, usually less than 1,000 m^2, and the latrines are typically made of bamboo or wood but occasionally may be made of brick.

Fish culture may also be carried out using fecally polluted surface waters. In Bojongloa, Bandung, there are more than 200 ha of ponds into which polluted surface waters are introduced. A similar system also exists in Bogor (Fig. 3), but in Cianjur fish are cultivated in raceways formed by screening a small, rapidly flowing, polluted stream which runs through the center of the city. In the cities of Cianjur, Bandung, Bogor (Fig. 4), and Sukabumi, fish are reared in

Fig. 3 Fish cultivation in ponds fed with fecally polluted surface water, Bogor, Indonesia.

Fig. 4 Fish cultivation in cages immersed in a fecally polluted stream, Bogor, Indonesia.

cages immersed in the polluted streams and rivers in the center of the cities. The cages are square or rectangular in shape with sides less than 3 m, and are constructed out of wood or bamboo strips with small spaces between to allow water to flow through the cage. They normally sit on the bottom of the stream or river bed and project slightly into the air, but are occasionally completely immersed in the flood season. The fish feed mainly on benthic invertebrates which develop in profusion in nutrient polluted water, but may ingest fecal solids.

Aquatic Macrophytes

The discussion above is restricted to fish culture. However, in several parts of Asia floating or creeping aquatic macrophytes are cultivated for human vegetables in water that may be intentionally or incidentally contaminated with fecal matter e.g., *Ipomoea aquatica*, water spinach, and *Neptunia oleracea*, water mimosa.

WET SYSTEMS

Despite the relatively low number of people in developing countries served by waterborne excreta collection systems or sewerage, there are several examples in the world of commercially viable fish culture systems using sewage. The country with the largest number and largest area of sewage fed fish pond is India, and more recently China has been developing sewage fish culture. Perhaps the best known system is the Munich sewage fish pond system in Germany. There is also sewage fish culture on a more limited scale in Hungary and Israel.

India

It has been estimated that there are more than 132 sewage fed fisheries in India, covering an area of about 12,000 ha, most of which are located in West Bengal. The most extensive is the Calcutta sewage fisheries (Fig. 5). Since the latter is the largest single excreta resource recovery system in the world, it is described in some detail.

The main sewers of Calcutta began to function in 1875, after 16 years of construction, and combined sewage and storm water was discharged through an outfall into the River Bidyadhari about 8 km to the east of the city. Since the river was tidal, the sewage was stored in large reservoirs for up to 8 hours, and emptied into the river when the water level went down sufficiently at ebb tide. In 1904 the first warnings were sounded that the sewage drainage outfall was seriously threatened, since the river has silted up only 2 miles downstream by about 10 m in the last 21 years; 8 years later in 1912, the river bed had silted up another 6 m. Attempts were made to revive the river by opening up spill areas and dredging, starting in 1918, but by 1928 it was declared impossible to maintain the River Bidyadhari.

The main reason for the silting up of the River Bidyadhari was the conversion of the "Salt Lakes," a vast area of waterlogged swamps that formed the spill area of the river, into salt water ponds. The spill area was almost 180 km^2 at first, but dwindled to about 80 km^2. At the turn of the century the swamp was divided by embankments into 80 to 400 ha areas called "Nona Bheris." Water was let in at high tide bringing sea bass, mullet and prawns, but at ebb tide when some of the water was let out, the organisms were retained in the ponds by bamboo screens. The salt water fish farming was initially very profitable since few costs were involved, but the restriction of river water movement led to a rapid silting up of the river since silt brought up on the high tide could not be spread over the spill area and was deposited on the river bed. Due to the increased siltation in the river, it gradually became more difficult to effect water exchange in the fish ponds. Furthermore, due to the increased volume of sewage from city expansion and the reduced cross sectional area of the river, water pollution increased which eventually reduced the growth of the fish. The once powerful river dwindled so much in cross section that it became a high level sewage channel and the once profitable "Nona Bheris" became undrainable swamp.

In 1930, a landowner discovered he could cultivate carp by letting in small doses of sewage to the swampy area. The results were so good that within a short time practically the whole area was converted into a fresh water aquaculture system. However, difficulty was experienced in draining the ponds until the storm water channel was constructed from Bantala to Kulti in 1940.

Calcutta sewage disposal became of great concern between 1920 and 1930 due to the deterioration of the River Bidyadhari. In 1935 it was decided to construct two channels, one for sewage and the other for storm water. City sewage is taken through a network of sewers and sewage channels to the eastern side of the city and is discharged into the dry weather flow (DWF) channel through a number of siphons installed at different marginal points. The DWF channel leads directly to two sedimentation tanks and associated sludge drying lagoons at Bantala, near the old outfall on the east bank of the now defunct River Bidyadhari, about 8 km from the city. The sedimentation tanks were built to avoid deposition of sewage solids along the 27 km DWF channel built to take clarified effluent to an outfall at Kulti, an estuarine tributary of the River Royamangal. It has been reported that as much as 60% of the water in the DWF channel was utilized for fish culture between February and May in some years. The two circular tanks, the Pruss type from Germany, were able to remove more than 85% of settleable sewage solids with a 1.5 hr retention time from Calcutta sewage. They were the largest of their kind in the world when constructed in 1943, with an internal tank diameter of 78 m, but are hopelessly inadequate to handle the volume of Calcutta sewage today. A storm water flow (SWF) channel was also built adjacent to the DWF channel and has an outfall at Kulti. However, for practical purposes the SWF channel is now another sewage channel since the DWF channel is insufficient to handle the increasing volume of sewage from the expanding city. When the sewage drainage canals were completed in 1940 it was considered that dilution at the Kulti outfall would be sufficient to deal with the clarified effluent without impairing the river quality. However, due to the current overloading of the system, much crude sewage is discharged into the river. Due to tidal action, the discharged sewage oscillates without dissipating quickly and a length of about 45 km of the estuarine creek is a vast septic tank.

The construction of the DWF and SWF channels has split the formed Bidyadhari spill area into two sections, called the North and South Salt Lakes. The new drainage channels provide an almost unique gravity feed facility to the fish farmers for adding sewage to and draining water from the ponds. The DWF channel carrying pre-settled effluent from the sedimentation tanks usually runs at a high level, and its level can also be adjusted by a regulator to ensure almost continuous feeding of the fish ponds. The SWF channel running parallel to the DWF channel is a low level canal, the level of which can be kept about 1 m lower than the adjacent fish ponds, even in the monsoon season, and is used for draining water from the fish ponds. When the system was constructed, simultaneous feeding and draining could be given only to the North Salt Lake fish ponds; however, recommendations were made to improve sewage feeding to the South Salt Lake ponds which were only getting the facility of efficient drainage at the time.

The areas of the North and South Lake sewage fisheries were originally reported to be 3,804 and 1,903 ha, respectively, a total of 5,734 ha. However, a sizeable part of the North Salt Lake area was lost due to the Government of West Bengal Salt Lake Reclamation Scheme for Calcutta City expansion. The total area may now be only about 2,500 ha. Recently, it has been reported that there are plans to improve the underdeveloped South Salt Lakes fishery to compensate for the loss in both fish supply and jobs of fishermen. Annual fish yields for the North and South Salt Lake fisheries were reported to be 5,222 and 1,849 tons/yr, respectively. Thus, the two areas supplied about 14.5 and 5 tons of fish/day to the Calcutta markets, respectively. The mean fish yields for the two areas were only 1,373 and 958 kg/ha/yr, respectively. These yields are low compared to those obtained in experimental ponds with good management and do indicate that there is potential for considerably increased yields from the current, reduced fishery area. There should be a financial incentive to improve the efficiency of sewage reuse, since fish on Calcutta markets fetch up to US$2-3/kg.

Traditionally, Indian major carps have been cultured in the system, but tilapia is now grown also. The ponds are dewatered for 1-2 months in February and March for removal of vegetation and mud. Paddy may sometimes be cultivated as an alternative crop. Water is fed into the ponds from adjoining canals by the end of March or the beginning of April to a depth of 15 cm. Sewage is then fed in to increase the average depth to about 90 cm, i.e., the initial ratio is 5:1, sewage:water. After 15-20 days the sediments have settled and the water turned greenish in color due to phytoplankton growth. After stocking the fish, sewage is subsequently introduced slowly about once/month over a period of 5-10 days, so that the proportion of sewage to water in the pond is about 1:4. Considerable skill and experience in feeding the ponds with raw sewage to avoid deoxygenation of the water has been acquired by the fish farmers. During the monsoon season, pond water must be drained off to prevent flooding. Fish attain marketable size in 5-6 months after which fish are harvested continuously until February when the ponds are emptied.

China

The predominant types of excreta collection and reuse in China are the "dry" systems, but sewerage systems are being installed in the centers of the larger cities. Fish cultivation in waste water began in 1957 but by 1964 it was reported that there were about 670 ha of wastewater fed fish ponds in 42 cities. The mean yields of wastewater fed fish ponds were reported to be about 3-4 times those of other ponds with more than a 50% saving in operating costs.

Fig. 5 Sewage fed fish ponds, North Salt Lake area, Calcutta, West Bengal, India.

Fig. 6 Sewage fed fish ponds, Munich, Federal Republic of Germany (photo courtesy Dr. jur. Wolfgang Kasser, Bavarian Electricity Generating Authority).

In Changsha, Hunan Province, there was a total of 160 ha of wastewater fish ponds in 1972 but by 1979 this had increased to 270 ha. The latter represents only about 25% of the total fishery area, but produces nearly 50% of the total fish yield of the city. There is an experimental pond of 7 ha at Chenjiahu which produced 70 tons of fish/yr i.e., a yield of 10 tons/ha/yr. In the Xianghu Commune Wastewater Fishery Station with 147 ha of water area, the mean fish yield is 6 tons/ha/yr.

In Changsha, dilute wastewater is introduced into ponds which are 0.7 m depth at the shallow end, increasing to 2.5-3 m depth at the deeper end. There are three sections: primary sedimentation, purification, and utilization, the interfaces between which change with the weather or wastewater flow. Since it is inconvenient to remove bottom deposits of mud, designs are now being studied with a centralized primary sedimentation pond with distribution of effluent to several fish ponds.

Two types of pond flow system are being used, a continuous system suitable for more stable flow and larger ponds at Chenjiahu, and an intermittent system suitable for unstable flow and smaller ponds at Xianghu. The detention time of the water in the ponds varies from 10-40 days, depending on the water quality. The ponds are dewatered every 2 years for desludging. More than 70-80% of the fish stocked are plankton feeding silver and bighead carp, with the remainder being bottom feeding crucian carp. The effluents are still quite high in total suspended solids and total nitrogen but with a detention time of 40 days, total suspended solids decreases from 228 to 16 mg/l, BOD_5 decreases from 301 to 18 mg/l, and dissolved oxygen increases from 0 to 6.5[5] mg/l, indicating a high degree of water purification.

Federal Republic of Germany

Sewage fish ponds were developed in Germany at the end of the nineteenth century to monitor the quality of drainage water from sewage irrigated fields used for sewage treatment. Following observations of fish culture using excreta in the Far East, the concept of combined sewage treatment and fish culture in a single pond was developed in Germany in the early twentieth century. Problems were initially experienced with rapid sludge build-up in ponds since raw sewage was used, but later mechanical treatment was employed before sewage recycling.

Sewage fish ponds were built in several locations in Germany, but the largest and best known is at Munich (Fig. 6). The sewerage system was built in Munich in 1881, and by 1900, 75% of the population of the city was connected. However, there was no sewage treatment until 1925 when an Imhoff tank was used for settling of solids and biogas production from the digestion of the solids. The sedimentation tanks retain 69% of the settleable solids and the clarified effluent is pumped to the sewage fish ponds. However, the raw sewage is relatively diluted due to large volumes of cooling water received from breweries. The 233 ha fish pond complex was built on moorland of little agricultural value. There is an adjacent 615 ha reservoir for water storage for hydroelectricity generation; the reservoir also receives the total wastewater flow when the ponds are dry during the winter.

The Munich sewage fish ponds started operation in 1929. They were designed to treat the sewage, following mechanical treatment, of 500,000 people, with a peak of 700,000 people, equivalent to a 2,000 population equivalent/ha. However, a higher sewage treatment population equivalent/ha should be possible in the tropics with higher temperatures. With the increase in population of Munich, the sewage fishery was unable to cope with the total volume of sewage, and the excess was discharged into the River Isar, which became increasingly polluted. The sedimentation plant was extended from 1957 to 1960 with the construction of a parallel group of settling basins and a heated digester. In addition, two activated sludge units were built in 1967 and a third one in 1974. However, the sewage fish pond complex still treats about 25% of the settled wastewater from Munich; from 1972 to 1975 a mean of 23.6% of the city's settled wastewater was treated. The mean organic loading to the 233 ha fish pond complex varied from a minimum of 7 to a maximum of 18 tons BOD_5/day between 1965 and 1975, equivalent to a mean yearly areal organic loading of 30 to 77 kg BOD_5/ha/day.

The fish pond complex consists of ponds varying from 0.3 to 10 ha in size, which are rectangular in shape, with a length to width ratio of 3:1. The mean depth is 0.9 m, but the ponds vary in depth from 0.5 m at the inlet, increasing to 1.5 to 2 m depth at the outlet. There are 30 large ponds, and a section of smaller associated ponds for breeding, nursery, overwintering and storage of harvested fish prior to marketing. Drainage and outflow of water is regulated at the outflow end of the large ponds. The pond bottom has trenches for emptying the pond and harvesting the fish.

The wastewater is introduced at the smallest side of the 30 larger ponds by 3 sprinklers/pond, which dilute it with river water at a ratio of sewage to river water 1:4 to 1:5. The wastewater is lifted up by 6 centrifugal pumps to a level of 10 m and is led to the pond system by iron pipes. The wastewater flows along the ponds in reinforced concrete pipes 2 m in diameter for 7.5 km. The river water is very cold so is introduced into 2 ponds at the western end of the fish pond complex to allow it to warm up and to sediment the silt load. The wastewater is sprayed through a nozzle in each sprinkler and falls 2 m to the pond surface above which there is a cascade of river water which effects immediate mixing of sewage and river water. The dilution rate of wastewater can be regulated separately for each pond. The detention time in the pond is normally 42 hours, with a minimum of 20 to 30 hours. Over many years there has been a mean of $1.65m^3$/sec of wastewater and $9.5m^3$/sec of river water, with maximum loadings of $2.5m^3$/sec of wastewater and $8\text{-}10m^3$/sec of river water.

Fish are cultivated for only 7 months, from April to October, because of low winter temperatures. The ponds are filled up with river water in March and in mid April fish are stocked. Wastewater is introduced into the ponds slowly to allow adaptation of the fish. The ponds are normally drained in October but sometimes are filled with water from October to December. A minimum dry period from December to March is sufficient to mineralize any sludge accumulation; there has been no sludge removal in 30 years of operation.

Common carp is the major cultivated species and feeds on bottom invertebrates such as chironomids and tubifex, but also on zooplankton. Fish yields are a mean of about 500 kg/ha/growing season. Two weeks before fish harvest, only fresh water is added to the ponds for depuration.

Hungary

There is a limited amount of sewage fish culture in Hungary but few data are available. Following primary sedimentation, 60 to 100 m^3 of sewage, with a maximum of 150 m^3, are sprayed daily on each hectare of pond surface (Fig. 7). With a water consumption of 200 l/capita/day this is equivalent to a range of 300 to 750 population equivalents/ha. There is no effluent since the sewage added serves to replace water lost by seepage and evaporation. Fish yields of 1,700 kg/ha/growing season have been reported with a polyculture of filter feeding Chinese carp and common carp stocked in a ratio of 4:1.

Fig. 7 Sewage fed fish pond in Hungary (photo courtesy Dr. Elek Woynarovich).

Israel

It has been estimated that there are about 50 to 100 ha of fish ponds in Israel that receive sewage, usually from relatively small rural communities, kibbutzim, producing 100-600 m^3 of wastewater/day. Since the sewage has a high BOD_5 of usually 250-300 mg/l there is no effluent from the ponds and the sewage replaces water lost by seepage and evaporation. A polyculture of common carp, Chinese carps, mullet and tilapia are normally reared. The range of organic loading is about 25-45 kg BOD_5/ha/day, assuming that the sewage added just balances the 1-1.5 cm daily fall in pond water level due to seepage and evaporation. Fish yields may be as high as 5,000 kg/ha/yr (extrapolated to a calendar year). The fish are depurated in clean pond water for several weeks at the end of the growing season to remove residual objectionable odors and pathogens. However, recently there has been a ban on using raw sewage in Israel for fish culture because of concern over possible health effects and only secondary sewage effluents may now be used.

CHAPTER III. AQUACULTURE IN RELATION TO SANITATION OPTIONS

There are various viable technology alternatives to provide the urban poor and rural communities in developing countries with adequate sanitation at a cost they can afford. World Bank studies have identified a range of appropriate technologies besides pit latrines and conventional sewerage, the two options generally only considered by sanitary engineers. In this chapter, the feasibility of linking aquaculture to the various sanitation technology options is discussed.

It is useful to split the overall process of the collection and disposal of human wastes into various stages: deposition, collection, transportation, treatment, reuse (which for the purpose of this report is aquaculture), and disposal. A major technical difference between the various sanitation options is the amount of water used, which leads to a major division into "dry" and "wet" sanitation systems; in all options except conventional sewerage, little flushing water is used and excreta are retained at least temporarily in a toilet receptacle or latrine onsite. Both the septic tank and vault are considered as "dry" systems in this report even though they may be connected to a cistern flush, since sludge or nightsoil, respectively, must be removed from the site by cartage. The pour flush water seal is regarded as a method of deposition and not as a type of toilet as in the World Bank Sanitation reports, since it may be connected to various kinds of excreta collection systems. Nightsoil or sludge collected in the receptacle must be periodically transported, which is referred to as cartage in the "dry" systems, but in conventional sewerage excreta is removed automatically by water flow in sewers. The degree of onsite treatment in the "dry systems" increases with the detention time in the receptacle but excreta treatment following cartage away from the site is required, and could involve reuse through aquaculture. More details of excreta disposal systems can be found in Kalbermatten, Julius, and Gunnerson (1980) and Feachem and Cairncross (1978).

It is technically feasible to link all the sanitation options with aquaculture. The different kinds of "dry" excreta, fresh or after various types and degrees of treatment, may be added to a pond for resource recovery (Fig. 8). Similarly, sewage may be recycled in aquatic systems (Fig. 9). The pond into which the various types of excreta may be added for reuse is referred to as a maturation-fish pond since in most cases adequate levels of dissolved oxygen need to be maintained for fish growth. From a biological point of view there is little difference between a maturation pond stocked with fish in a series of stabilization ponds and a fish pond fertilized with various types of excreta.

A brief discussion is presented below of the feasibility of excreta reuse in aquaculture in association with various sanitation options.

COMPOSTING TOILET

There are two types of composting toilets, continuous and batch. Continuous composting toilets are extremely sensitive to the degree of user care and even if

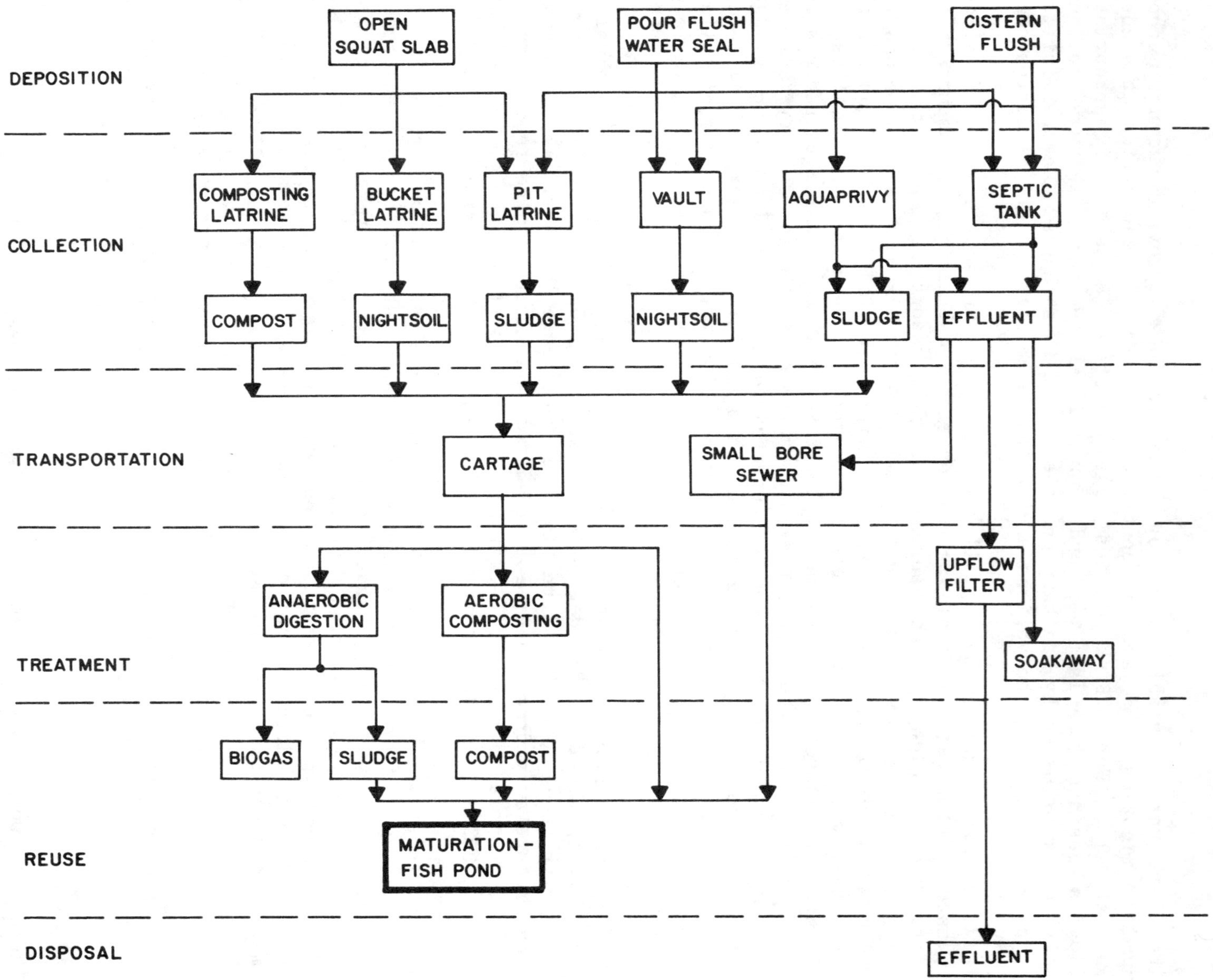

Fig. 8 The relationship between "dry" sanitation technology options and aquaculture.

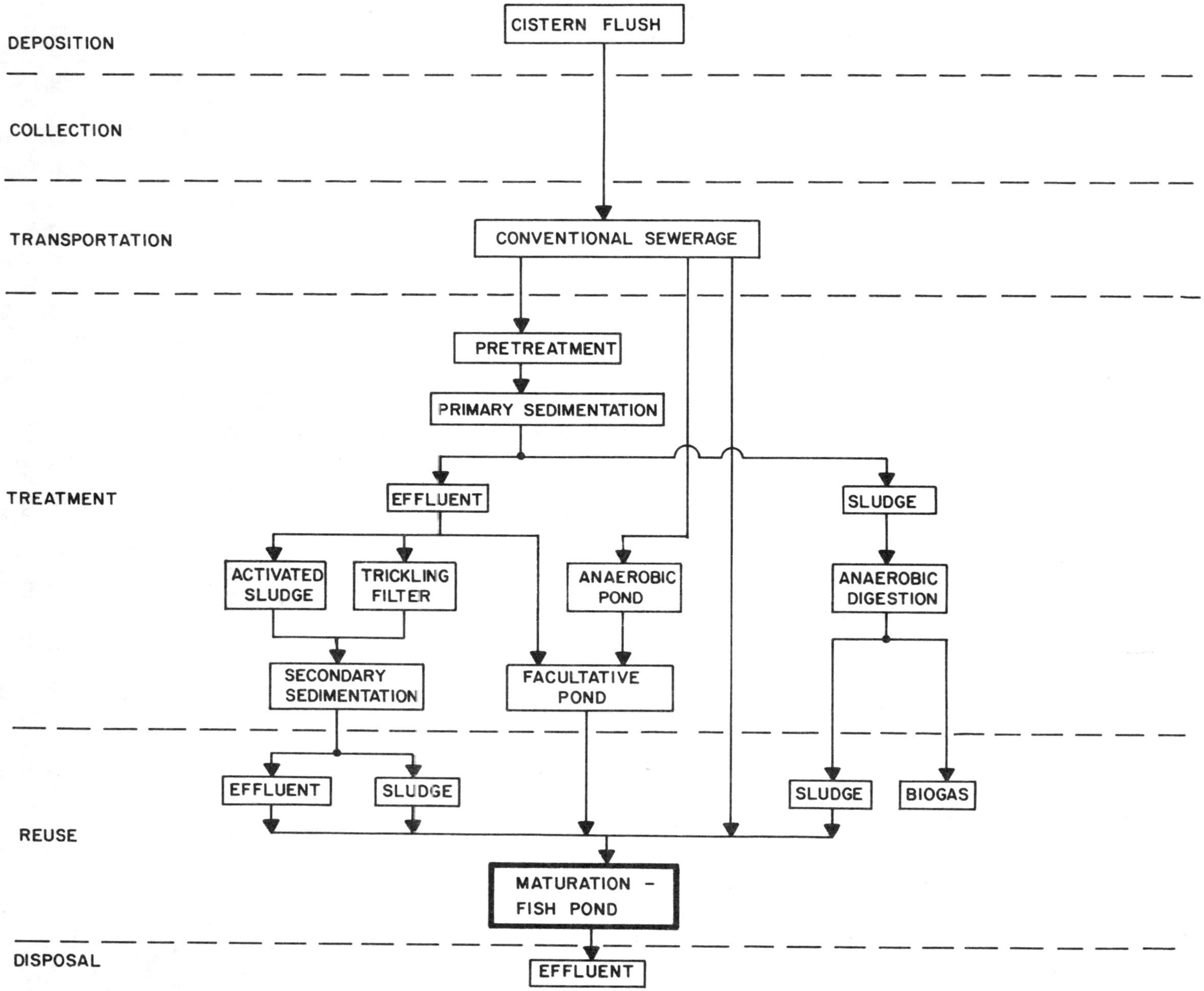

Fig. 9 The relationship between "wet" sanitation technology options and aquaculture.

used properly, fresh excreta may slide into the pile of composted waste and cause a health hazard if recycled without further treatment. The continuous composting toilet is not recommended. However, the double vault composting (DVC) toilet, the most common type of batch composting toilet, is safe if the compost is stored for 12 months prior to use. A 3 month retention time would produce compost free of all pathogens except the more persistent helminth ova such as Ascaris. The DVC is reported to be widely used in the rural areas of Vietnam and the compost added to fish ponds.

BUCKET LATRINE

The operation of bucket latrines is normally insanitary and it is difficult in practice to ensure that improved bucket latrine systems are operated satisfactorily in developing countries. Thus even an improved bucket latrine system cannot be recommended for new installations (Kalbermatten, Julius, and Gunnerson, 1980). However, a wooden bucket collection system is widely used in urban and rural China; the buckets are tipped into sealed carts or tanks and sanitary conditions are apparently maintained by a high degree of user care and motivation. Excreta collected in this way are used in aquaculture in China, reportedly after a period of storage in sealed tanks.

PIT LATRINE

This is likely to remain the most common technology, particularly in rural areas, due to its simplicity. An improved version is available which should lead to more widespread acceptance, the ventilated improved pit (VIP) latrine. Pit latrines can be desludged and the excreta recycled in aquaculture, preferably after the pit has been sealed and not used for a minimum of 12 months when only a few Ascaris ova at most would be viable. Pit emptying could be done by householders or by municipal or private concerns who presumably could sell the contents as an input for fish ponds.

VAULT

Nightsoil is stored in containers located in or near the house and is periodically removed. Such cartage systems for excreta removal are widespread in urban areas in Asia, particularly in Japan and Taiwan, China; large areas of Tokyo are serviced by vacuum trucks and conservancy vaults. Vaults can be hygienic with a pour flush water seal or low volume cistern flush and although emptying the vaults by hand using dippers, buckets and open carts is insanitary, vacuum trucks may be used. The frequency of excreta removal depends on the size of the vault, from daily to every two to four weeks. Thus, although nightsoil cartage and treatment systems tend to have higher health risks than sewerage, these can be considerably reduced by the use of well designed systems such as in Japan.

There are several advantages for the vault and vacuum truck sanitation option. The water seal bowl and vault can both be manufactured locally and sold as a market commodity. A minimal amount of water, about 2-6 liters per capita per day is required. The option is suitable for high population density urban areas since the excreta can be removed readily by vacuum carts and trucks. Such a collection system is highly flexible, an important consideration in rapidly expanding and changing urban areas in developing countries. Furthermore, the vault system can be readily upgraded by conversion to cistern flush septic tanks with upflow filters or small bore sewers, with a reduction in the frequency of the removal of the vault contents to once every few months. A key problem with the system is the organization of excreta removal since there is the fundamental need to remove nightsoil regularly. However, the vault is the best sanitation technology from the point of view of aquaculture reuse since there is the greatest degree of conservancy of excreta among the various sanitation options.

AQUAPRIVY AND SEPTIC TANK

In both these sanitation technologies there is a separation of solid matter by sedimentation from a liquid effluent which flows out of the tank to soakaways (pits) or tile drain fields. The septic tank is generally considered to be a sewage system since excreta is mixed with large volumes of water from a cistern flush and is carried by a short sewer to the septic tank located underground, away from the house. Since sludge that accumulates in the tank must be periodically removed, the septic tank is best regarded as a "dry" system from a reuse point of view.

Many cities in Asia rely on septic tanks, which can lead to pollution of the surrounding area from either insufficient land for effluent disposal or impermeable soil. Generally cesspools are installed with outlet pipes, which may be unlined initially so that the surrounding soil may act as a soakaway, but in dense urban areas the soil pores soon become blocked, and it then functions like a septic tank. To safeguard the environment from pollution, the "septic tanks" in Asian cities could be equipped with anaerobic upflow filters to further treat the effluent, or converted to watertight vaults with more frequent removal of the contents. The latter option would of course provide more excreta for reuse in aquaculture. The effluent of aquaprivies and septic tanks could be removed by small bore sewers which could be led to a maturation-fish pond.

CONVENTIONAL SEWERAGE

The excreta is mixed with large volumes of water from a cistern flush and is removed from the house in a flow along sewers, normally to a treatment plant. The major advantages of conventional sewerage, in addition to safeguarding public health, are a high degree of user convenience. However, there are numerous technical and financial constraints which make it an unsuitable option for developing countries.

Following preliminary treatment to remove large floating objects and sand and grit, suspended sewage solids are removed by primary sedimentation in sedimentation tanks. The sludge is hazardous to health and highly odorous when fresh and should be treated before reuse by digestion. The three major types of secondary sewage treatment are trickling filters, activated sludge, and stabilization ponds. The first two are the least suitable for developing countries since they are expensive and require a high degree of maintenance. In addition, the effluents may still be hazardous from a health point of view, and contain significant amounts of nutrients. There are reports of effluents from both trickling filters and activated sludge plants being further treated in maturation ponds and used for aquaculture following secondary sedimentation. Sewage sludges have also been used for the experimental feeding of fish.

Stabilization ponds are recommended for treating sewage in the tropics since they are cheaper to build and maintain than trickling filters and activated sludge plants and are vastly superior from a public health point of view. A properly designed series of ponds with a minimum detention time of 25 days can produce a final effluent with no protozoa and helminth ova and only a very low survival rate of bacteria and viruses. A series of conventional stabilization ponds normally consists of an anaerobic pond, which functions like a septic tank to sediment sewage solids, followed by facultative and maturation ponds. These latter two contain large densities of phytoplankton suitable for feeding fish, but the dissolved oxygen regime is normally suitable for aquaculture only in maturation ponds. From a sanitary engineering point of view, anaerobic and facultative ponds are designed for 5-day biochemical oxygen demand (BOD_5) removal and maturation ponds for pathogen removal, but the latter may be suitable for aquaculture. As indicated in Fig. 9 there are various options for recycling sewage in a series of stabilization ponds, which depend to a large extent on the strength of the raw sewage. If the sewage is weak or is diluted considerably, it may be added directly to the maturation-fish pond. Dilution may be achieved by mixing with unpolluted water or by adding only small amounts of raw sewage to the pond so that the pond water itself is the major source of dilution water. Primary sedimentation is normally used to remove sewage solids in both waste treatment and aquaculture reuse, otherwise they would lead to a rapid sludge build-up in the ponds.

There are technical constraints to recycling nutrients contained in excreta in the form of sewage. It is difficult to prevent the contamination of domestic sewage with toxic chemicals such as heavy metals and synthetic organics from domestic nonfecal substances or from industrial sewage. These may accumulate in aquatic organisms and render them unsuitable as sources of human or animal feed. Toxic chemicals may be present in such concentrations in sewage that they even constitute a threat to fish growth in maturation ponds. Hard detergents in sewage from Haifa, Israel have been reported at levels of 17 mg/l ABS, while the lethal concentration for carp is 10 mg/l; it was also found that sublethal concentrations of ABS detergents reduced the growth rate of fish.

A further constraint to the use of sewage for aquaculture is the loss of nutrients in the final effluent. Since sewage has a high water content, a flow through system must be used to treat and reuse the excreta, but this leads to poor resource recovery. The amount of effluent could be reduced by increasing the detention time of the system but this would lead to a greatly increased land requirement which would seldom be feasible due to high land costs in suburban areas of cities where sewage would be available for treatment. Furthermore, as the detention time of maturation ponds is increased, the concentration of plankton in the water decreases, with lower fish yields per unit area of pond surface.

SUMMARY

Since "dry" forms of excreta have a much lower water content than sewage, water will need to be added to maturation-fish ponds used to recycle them to balance water losses due to seepage and evaporation. Thus, there should be no effluent from ponds used to recycle "dry" forms of excreta .

An important result of the World Bank study is the development of the concept of "sanitation sequences," step-by-step improvements which can be implemented as the socio-economic status of a given community increases. A significant point is that none of the sanitation sequences leads to conventional sewerage and, even in urban areas, the final upgrading is generally to a low volume cistern flush toilet connected to a vault that overflows into a small bore sewer. The drawbacks to the use of sewage in aquaculture discussed above also lead to a similar conclusion: that "dry" or conservancy systems of excreta collection and treatment are better than "wet" or conventional sewerage for aquacultural reuse.

CHAPTER IV. BIOLOGICAL CONSIDERATIONS IN AQUACULTURE EXCRETA REUSE SYSTEMS

A major constraint to the development of appropriate sanitation technology incorporating waste recycling is that two widely separated disciplines are involved which have had little interaction in the past: sanitary engineering and aquaculture. Waste treatment involves using naturally occurring microorganisms to degrade and stabilize organic waste, while fertilization of a fish pond with organic wastes involves the stimulation of the growth of natural biota, especially microorganisms, as fish food. Clearly, the biological processes involved are similar, and it is feasible to incorporate resource recovery into waste treatment systems.

SANITARY ENGINEERING CONCEPTS AND AQUACULTURE

The theoretical basis of organic waste application to water has been developed by sanitary engineers for waste treatment, but has relevance for the design of human waste recycling schemes. Biological oxygen demand (BOD) and chemical oxygen demand (COD) are used to measure the oxygen required to oxidize organic wastes. Biological oxygen demand is usually measured over a period of 5 days at 20°C (BOD_5), but this has limited value for aquaculture, particularly in the tropics. A more appropriate test for aquaculture is overnight BOD of pond water at pond temperature ($BOD_{0.5}$) since the most critical factor in waste loaded ponds containing fish is the decrease of dissolved oxygen in the pond water during the night. As explained below, the major cause of this is the respiration of the pond biota, and not the oxygen demand of wastes added to the pond.

The organic loading, the rate at which biodegradable organic matter is added to a pond, is measured in kg BOD_5 or COD/ha/day. High organic loadings are employed by sanitary engineers since one of the objectives of waste treatment is to utilize the minimum amount of land possible, to reduce waste treatment costs. If fish are to be grown in waste loaded ponds, then lower organic loadings must be used. The principle in waste loaded fish ponds is to add enough waste to provide adequate nutrition for the pond biota which the fish consume, but not enough to lead to dangerously low dissolved oxygen levels at night which could endanger fish survival. The addition of waste to fish ponds is still very much an art learned through experience, but research is being carried out to quantify organic loading of excreta in aquaculture systems. Such data are needed to provide bioengineering design criteria for constructing excreta reuse sytems.

DISSOLVED OXYGEN IN WASTE LOADED PONDS

Since most of the pond biota, particularly fish, require oxygen for respiration, it is essential that adequate levels of dissolved oxygen (DO) be maintained in fish ponds. Biological activity in the pond accounts for the greatest variations in DO. The major source of DO in the water is the photosynthesis of phytoplankton during the daylight hours, while the major cause of oxygen depletion is the respiration of pond biota. In fertile ponds loaded with organic matter, there are large diurnal fluctuations in DO. In a well managed waste loaded fish pond, the minimum DO in the early morning hours is only a few mg DO/l but is supersaturated with oxygen in the late afternoon. The following equation may be used to estimate the DO at dawn, the critical period in a waste fed fish pond:

$$DO_{dn} = DO_{dk} \pm DO_{df} - DO_m - DO_f - DO_p$$

where

DO_{dn} = DO concentration at dawn
DO_{dk} = DO concentration at dusk
DO_{df} = DO gain or loss due to diffusion
DO_m = DO consumed by mud
DO_f = DO consumed by fish
DO_p = DO consumed by plankton

Waste loaded fish ponds are normally supersaturated with DO at dusk due to photosynthetic oxygen production. It has been estimated that the net loss of oxygen from the pond by diffusion in a pond with double supersaturation in the late afternoon may exceed 4 mg DO/l, although there may be diffusion into the pond from the air during the early morning hours. Mud respiration probably lowers the DO in the water by less than 1 mg/l overnight. A fish population weighing 3,000 kg/ha would also lower the DO in the water by about 1 mg/l overnight. However, the most important factor in reducing the DO of the water overnight is the respiration of plankton (backterioplankton, phytoplankton, and zooplankton). It has been estimated the plankton respiration can lower overnight DO by 8-10 mg/l, much more than the other factors involved. Thus, the most important aspect concerning oxygen depletion in waste loaded ponds is the development of dense blooms of phytoplankton.

The oxygen demand of organic waste is not included in the above equation, because it is accounted for in the BOD of pond water and pond mud. However, it has been estimated that the BOD of excreta added to a fish pond is a minor factor in the nighttime consumption of oxygen, less than 0.5 mg/l. The oxygen demand of organic waste added to a pond is a major factor only in an anaerobic pond without phytoplankton, in which a high organic loading leads to depletion of oxygen by bacterial respiration.

Thus, it is not the BOD of the excreta itself that causes the greatest reduction of dissolved oxygen in a waste loaded fish pond, but the respiration of the phytoplankton that develops as a result of the nutrients contained in the excreta. The significance of the effect of adding different forms of excreta with varying BOD_5 and nutrient characteristics to fish ponds is currently being assessed.

Due to poor knowledge concerning rates of nutrient uptake by pond biota, and losses of nutrients from the system by physical means such as precipitation and volatilization, it is difficult to establish rates of excreta addition. However, since the nutrients are taken up by bacteria and plants (phytoplankton or macrophytes) the excreta should be distributed as uniformly as possible throughout the water. To obtain more effective nutrient uptake, excreta is best applied frequently, in small doses. The principle of adding sewage to fish ponds in small doses at frequent intervals to prevent deoxygenation was well known in Germany by the early 1930's.

In a waste recyling system using "dry" excreta, it would be ideal to add the waste daily, but for practical purposes to reduce labor costs, it may be feasible to load twice a week or even weekly. Excreta could be distributed manually using buckets, or by gravity or pumping through channels or pipes of brick or concrete to various locations in small ponds. In larger ponds, boats may be required to distribute "dry" excreta. The use of a spray or sprinkler system is not recommended since aerosols may distribute pathogenic microorganisms up to about 1 km from the source of excreta distribution.

NUTRIENTS IN WASTE LOADED PONDS

The major reason for adding excreta to an aquatic reuse system is to provide substrates for bacteria, the breakdown of which releases inorganic nutrients for plant growth at the base of the food chain. Thus, the nutrient content of the waste is its most important characteristic since the nutrients eventually become constituents of the desired cultivated organism in the system. For the design of waste loaded aquaculture systems, a knowledge of the major nutrients required for the growth of bacteria and phytoplankton or other plants is needed. Phytoplankton have a similar chemical composition when grown in a nutrient rich medium in which light is not limiting: about 50% carbon, 10% nitrogen, and 1% phosphorus. Thus, the major nutrient in an aquatic system is carbon, which is not normally considered in crop fertilization because it is provided by the air. Since potassium has not been shown to be limiting in fish growth, the major nutrients in aquatic systems are nitrogen-phosphorus-carbon (NPC), not nitrogen-phosphorus-potassium (NPK) as for land crops.

BIOTA OF WASTE LOADED SYSTEMS

There is a wide variety of organisms that can be cultivated in excreta loaded systems. A brief discussion of the major types of organisms is presented, with comments on their suitability for potential commercial excreta resource recovery systems in developing countries.

Phytoplankton

A major characteristic of excreta loaded systems is the occurrence of dense growths of phytoplankton. In waste stabilization ponds there is a symbiotic relationship betwen bacteria and phytoplankton; the former degrade the organic wastes to release inorganic nutrients which are assimilated by phytoplankton, and the oxygen required for bacterial respiration is provided by the photosynthesis of the latter. A major concern of using stabilization ponds to treat organic wastes is the high content of phytoplankton in the effluent, which is considered to be a form of pollution for the receiving water. However, since phytoplankton consist of about 50% protein on a dry weight basin, engineers have sought to harvest them as a potential source of animal feed, and more recently as a source of chemicals such as various lipids, glycerol, and natural pigments.

A most significant development of the concept of utilizing waste grown phytoplankton was the invention of the high rate stabilization pond by Oswald and co-workers in California (Shelef & Soeder, 1980). By treating sewage in shallow ponds with mechanical mixing, it is possible to obtain extrapolated phytoplankton yields greater than 100 tons dry weight per hectare per year (t/ha/yr) in tropical climates. This would provide a protein yield from 10 to 50 times greater than soybean, the most prolific agricultural protein producer. However, a major problem is the separation of the phytoplankton from the water in the form of a dried, stable product. Despite 30 years of research using a variety of techniques - centrifugation, chemical flocculation, autoflocculation, electrifloatation, and microstraining - harvesting and processing the phytoplankton remains an expensive and relatively technically complicated process. During a project in Singapore, it was possible to remove the phytoplankton from the water by using a microstraining technique with a mesh size as small as 5 μm, but processing still remains a problem since the slurry with a solids content of only about 10% rapidly decomposes in the tropics. It may be feasible to feed the harvested phytoplankton in the form of a wet slurry to livestock, but there may be a need to rupture the cell walls of the algae to permit an acceptable level of digestibility; the latter has generally been achieved by centrifugation to reduce the water content followed by drum drying, which is costly.

A project was carried out in Thailand to investigate the possibility of using phytoplankton filtering fish to remove the phytoplankton from the effluent of a sewage driven high rate stabilization pond which was pumped to fish ponds. It was concluded that the system would probably not be economically viable since algal production and harvesting were out of phase; although the fish filtered phytoplankton and grew well, they were unable to remove algae as efficiently as they were produced by the high rate pond, and significant amounts of algae passed through the system in the effluent. Phytoplankton losses in the effluent could be reduced by increasing the detention time of the fish ponds, which would lead to greatly increased reuse efficiency in terms of fish yield per unit of algal production; but increasing the detention time of the fish pond would lead to a lowering of the algal concentration in the water and thus lower fish yields per unit area of fish pond, which would increase the area of land needed to reuse the phytoplankton.

The high rate stabilization pond still has great potential as an excreta reuse system, but in the absence of simple and economic techniques for harvesting and processing the phytoplankton, it cannot at present be recommended for implementation as a low cost sanitation technology for developing countries. The best strategy at present is to harvest phytoplankton by means of organisms higher up the food chain.

Macrophytes

Aquatic macrophytes have been inadvertently used to treat excreta discharged into the environment since time immemorial, although some are cultivated in tropical developing countries in waste loaded systems. However, certain types, particularly the water hyacinth, which causes major aquatic weed infestations throughout the tropics where water is present, comprises a widespread, unplanned waste treatment system (Fig. 10). Aquatic macrophytes are usually divided into life forms, the major types being: (i) emergent species which are rooted in shallow water with their vegetative parts emerging above the water; (ii) submersed species which are usually rooted and are predominantly submerged; and (iii) floating species with roots, if present, hanging in the water.

Fig. 10 Natural water hyacinth excreta treatment system, Sakorn Nakon, Thailand.

In the West, sewage has been added to wetlands with emergent macrophytes for treatment and disposal in certain areas for many years, but recently there has been a surge of interest in their use in sewage treatment because of the potential for reducing energy requirements, and operation and maintenance costs for wastewater treatment. The pioneering studies on the construction of artificial marshes of emergent aquatic macrophytes such as bullrushes, cattails, and reeds were conducted in Germany and later in the Netherlands, but the major focus of activity has shifted to the USA, where natural as well as artificial marshes are being used for the treatment of secondary sewage effluent. Since wetlands are suitable only for the treatment of dilute wastewater, and there is little commercial demand for the vegetation produced, these systems have little relevance for tropical developing countries.

Submersed macrophytes are not suitable for excreta reuse systems due to shading effects from phytoplankton which develop in fertile water and effectively eliminate them. Heavy growth of submersed macrophytes and phytoplankton are rarely compatible in the same water body.

Floating macrophytes have potential for waste reuse. Recently there has been much work in the USA on the use of floating macrophytes to reduce the concentration of phytoplankton in the effluent from waste stabilization ponds; the underlying principle is the ability of floating macrophytes, usually water hyacinth and to a lesser extent duckweed, to eliminate phytoplankton from the water column by shading and to take up the nutrients released by phytoplankton decay, thus clarifying the effluent. Floating macrophytes have great potential for use in excreta resource recovery systems in developing countries using "dry" excreta in single pond systems, without prior treatment associated with phytoplankton production. The two major species are the water hyacinth, and the duckweed <u>Spirodela.</u>

The water hyacinth is one of the fastest growing plants in the world. Most of the studies on its growth rate have been conducted in the southern USA, but extrapolations from summer growth rates, which approximate tropical conditions, give yields of over 100 tons dry matter/ha/yr in nutrient rich waters. The main problem with water hyacinth is the cost and labor involved in harvesting, transporting and processing it, since it is more than 90% water and has a low density of only 80 Kg/m^3. To be successful in an excreta reuse system, a revenue generating use for it must be found. There is potential for using water hyacinth in biogas production. It is unfortunate that fish do not consume water hyacinth directly, but there are various options for the use of water hyacinth in fish culture.

Duckweeds have good potential for use in excreta reuse sytems since they have several desirable attributes: a high growth rate, perhaps as high as 20 tons dry wt/ha/yr in a well managed system in the tropics; a crude protein content greater than 30% when cultivated in fertile water, which is only slightly less than that of soybean. They are readily consumed by a variety of domestic animals such as cattle, poultry, and herbivorous fish. The use of duckweeds in excreta recycling systems is one of the most promising options available. Until recently duckweed was commercially cultivated in <u>Taiwan</u>, China.

Although the above discussion concerns freshwater aquatic macrophytes, it may be feasible to develop an excreta recycling system with marine aquatic macrophytes or seaweed. In the Wood's Hole waste recycling system utilizing secondary treated sewage effluent mixed with seawater, a final polishing step using several species of red algae was added to remove nutrients from animal culture effluents. Subsequently, a one stage system was set up in Florida using seaweed alone, since the preliminary results were so promising. Seawater enriched with secondary treated effluent produced extrapolated annual yields under the best growing conditions of more than 100 tons dry weight/ha of the agar producing seaweed <u>Gracilaria.</u> This could be promising from a financial aspect since the price of good quality <u>Gracilaria</u> exceeds US$500/ton in Japan, and also from a public health point of view since the seaweed would be used for agar extraction. However, the economics of the system appear marginal at best due to large energy inputs in the flowing water system in the form of pumped water and air. With a less energy intensive system such as the one used in Taiwan, yields are about ten times less but the system is commercially viable. Great skill is required to operate such a system using excreta since rapidly growing green algal epiphytes and phytoplankton tend to develop and shade the red seaweed crop. Additional information on the fertilization and management of the Chinese *Gracilaria* cultivation system is required before an assessment can be made of the feasibility of seaweed cultivation using human waste.

Zooplankton

Zooplankton feed mainly on phytoplankton or on small bacterioplankton aggregates. A few years ago they were studied in the USA as a possible method of removing phytoplankton from stabilization pond effluents, but floating aquatic macrophytes are now considered to be more effective. The growth of zooplankton is unpredictable and thus difficult to control. Furthermore, the growth of zooplankton tends to be inhibited in ponds with high densities of phytoplankton, possibly due to elevated pH, often exceeding pH 9.5, caused by algal growth. A commercial application of zooplankton cultivation is as food for aquarium fish, which is hardly relevant for poor communities in developing countries. There is also considerable research into the cultivation of zooplankton for feeding to fry in fish, shrimp, and shellfish nurseries, generally with monocultures of phytoplankton, which also has little relevance to low cost sanitation technology.

Benthic Invertebrates

Probably the most important benthic or bottom living invertebrates in waste loaded ponds are the filter feeding larvae of midges, called chironomids or blood worms. They may exist in tens of thousands/m^2 in pond sediments with low dissolved oxygen due to the microbial decomposition of organic matter. Since they have a high protein content they are frequently collected as feed for aquarium fish by sifting through the bottom sediments with sieves. Like zooplankton, their cultivation is hardly relevant on a large scale due to the limited market for feed for aquarium fish.

Fin Fish

A wide range of fish species is reported to have been cultivated in excreta loaded systems: common carp, crucian carp, Indian major carps, Chinese carps, mullet, milkfish, catfish, tilapia, trout, and salmon.

The main characteristics of a waste loaded system is the predominance of phytoplankton, which often fluctuate in density and may on occasion develop high concentrations or "blooms." The most appropriate fish to rear in such a system are those which can feed on phytoplankton. The advantage of using a phytoplankton feeding fish can be seen by comparing fish yields from the Munich sewage fishery in Germany, about 500 kg/ha/season of the bottom feeding common carp, to a yield of 1,700 kg/ha/season in a sewage fed fish pond in Hungary in which a Chinese filter feeding fish was the main species.

There is still controversy over the role of phytoplankton as fish food and there are reports that fish cannot derive nutrition from certain phytoplankton since they may be toxic, too small to remove from the water, or indigestible. Certain genera of blue-green algae produce toxins which adversely affect livestock and water fowl, particularly in the midwestern USA. Although some of the same genera are common in waste loaded fish ponds in Asia, they do not appear to be toxic to fish. Most of the phytoplankton in waste loaded ponds are nannoplankton which are minute, usually less than 20 um. Phytoplankton feeding tilapias remove the algae from the water by entrapping them in copious secretions of mucus produced in the fish mouth and pharynx. Silver carp, which feed mainly on photoplankton, are able to strain the large cells out of the water through their gill rakers; there is evidence that small phytoplankton may be entrapped in mucus secreted by the labyrinthi-form organ. For many years it was believed that many of the most important phytoplankton in fertile ponds, blue green algae and euglenoids, were not digested by fish. It is now known that tilapia secrete acid, which lowers the stomach pH to less than 2 and occasionally to as low as pH 1-1.5, and this lyses blue green algae. The digestive mechanism of carps is less well understood since the stomach pH does not fall as low as in tilapia. There are conflicting reports concerning the ability of silver carp to digest green, euglenoid, and blue-green phytoplankton, but the fish certainly grow well in waste loaded ponds in which these algae predominate.

The most appropriate fish to rear in a waste fed pond are those which are tolerant to low levels of dissolved oxygen, which may be caused by unpredictable algal blooms. Fish species vary in their tolerance to low levels of dissolved oxygen. Air breathing fish such as the catfish Clarias are least affected since they can obtain oxygen from the air. Most species depend on dissolved oxygen in water. The most sensitive fish are the cold water salmonids, but these have little relevance for the tropics. Among the carps there are considerable variations, with common carp being much more resistant to low dissolved oxygen than silver carp. Tilapias are among the most resistant to low dissolved oxygen. However, data in the literature concerning the tolerance of fish to low dissolved oxygen have little relevance to natural situations with widely fluctuating levels since they were mainly derived from experiments with constant dissolved oxygen concentrations. Fish may be much more resistant to low dissolved oxygen concentrations for a short period of time than is generally realized. However, since there are several reports in the literature of silver carp mortality due to low dissolved oxygen, it may be wise to stock a more tolerant fish such as tilapia in an excreta loaded system.

Yields of fin fish in excreta loaded fish pond systems in the tropics should be at least 5-6 tons/ha/yr, and with good management could be as high as 10-12 tons/ha/yr. However, high yields are generally based on extrapolations from small experimental areas, so perhaps a yield of about 6-8 tons/ha/yr would be more realistic for a system with commercial size ponds.

Shrimp and Prawns

Penaeid shrimp and prawns are omnivorous and eat seaweeds, invertebrates, and detritus. In Asia they are often grown in brackish water milkfish ponds, alone or in combination with milkfish, since they have a high market value. There have been experiments in the USA and South Africa in which these crustacea have been cultivated in ponds receiving secondary sewage effluents, but mortalities occurred due to low dissolved oxygen levels at night caused by phytoplankton blooms. Good growth occurred if the water was aerated or if the organisms were cultivated in running water systems.

Since shrimps and prawns are sensitive to low levels of dissolved oxygen and have a high market value, it is illogical to grow them in excreta fed systems. It would be better to use pathogen free supplementary feed and not risk either mortality during cultivation or contamination of the final product.

Molluscs

Shellfish would appear to be ideal candidates for excreta fed recycling systems because they feed at the base of the food chain by filtering phytoplankton from the water. However, there are constraints to their use in waste resource recovery. They accumulate heavy metals, organic chemicals and pathogenic organisms and so could constitute a health hazard if used directly for human food or even animal feed. In shellfish cultivated in the Wood's Hole recycling project, heavy metals and organic trace contaminants were no higher than those in naturally unpolluted populations, and depuration of the bivalves for 10-14 days reduced any accumulated human viruses to undetectable levels. However, at Wood's Hole, secondary treated sewage effluent was used which would presumably contain fewer pathogenic organisms than normally found in excreta in a developing country.

Problems were experienced with the cold water shellfish cultivated at Wood's Hole, which tended to feed on specific algae only. Since it was not possible to control the species of phytoplankton in the algae ponds, the shellfish were sometimes unable to feed on the dominant genus present at a particular time, and were undernourished. This problem may not occur in the tropics, since warm water shellfish appear to feed on a wide range of algae and detrital particles.

A major constraint to the cultivation of shellfish in a waste recycling system is the need to provide water movement for oxygen supply, metabolite removal, and to provide food for the sedentary organisms. Attempts to grow oysters and mussels in floating racks in static ponds in South Africa led to high mortalities and lack of growth. Furthermore, the shellfish would need to be suspended throughout the water column for more effective filtration of phytoplankton, either on a bamboo framework or ropes suspended from rafts.

To use the shellfish, the meat would need to be removed from the shell, either by hand or mechanically, which would further increase production costs.

Shell fish do not appear to be suitable organisms for excreta reuse systems.

CHAPTER V. PUBLIC HEALTH ASPECTS

There are potential threats to public health from excreta reuse in aquaculture, which must be avoided. Systems need to be developed that do not pose unacceptable risks to health. From the four major groups of pathogenic organisms: viruses, bacteria, protozoa and parasitic worms, there are more than 50 infections (excluding different numbered types of viruses and serotypes of enteric bacteria) caused by lack of sanitation, although not all can be transmitted by improperly designed and managed aquaculture waste recycling systems. For a detailed account of health aspects associated with excreta, see Feachem *et al*. (1981).

HEALTH HAZARDS

There are various kinds of health hazards associated with excreta reuse.

Occupational Hazards

There may be an occupational hazard to those who are employed to work with excreta collection and reuse, but there is little epidemiological evidence. Workers may accidentally swallow pathogens or carry them home on their clothing or bodies.

A specific occupational hazard of excreta use could be schistosomiasis, but only in areas where the disease is endemic and where intermediate snail hosts are present in the ponds. This helminth disease, particularly *Schistosoma japonicum*, has been related to excreta reuse. Eggs may survive in feces for more than 1 week so that if fresh excreta is applied to ponds containing certain amphibious snail hosts, the latter may become infected. Larvae are shed into the water following development within the snails and can bore into human skin to infect pond workers. The avoidance of using fresh excreta can control the disease in areas where the helminth is endemic; storage of excreta for two weeks would render it free of eggs.

Spray distribution of sewage effluent could be a hazard to workers or people living near the reuse site. Aerosol droplets containing enteric bacteria are reported to travel up to 1.2 km and bacteria may be more infective when inhaled than ingested. There is also the possibility of infection by enteric viruses by inhalation.

Consumption of Contaminated Organisms

The degree of risk varies considerably with the type of pathogen conerned. It should be stressed that although there is little danger of disease from eating well cooked fish or vegetables since the heat destroys pathogens, the consumption of raw, partially cooked, or improperly preserved products, for example, poorly fermented fish which is widely consumed in parts of Asia, can be a serious health hazard. Perhaps the most significant health hazard, which is generally overlooked, is the danger from handling and preparing contaminated products.

Fish do not apparently suffer from infections caused by enteric bacteria and viruses that cause disease in warm blooded animals - humans and livestock - but they may carry pathogens passively. It is generally believed that fish carry human pathogens passively only in their intestines and on their body surfaces. However, recent work in Israel has revealed the rather startling fact that both enteric bacteria and viruses are able to penetrate various fish tissues (Buras *et al.*, 1982). The concentration of microorganisms in the water determines their presence in fish tissues, and there appears to be a threshold concentration in pond water below which microorganisms do not penetrate into fish muscle. The threshold concentration for viruses appears to be an order of magnitude less than for bacteria. The intraperitoneal fluid had even higher levels of bacteria and viruses, an important finding from a public health point of view since the fluid comes into direct contact with the handler when the fish is gutted, creating a potential source of infection.

There is little information on the passive transfer of protozoan cysts and helminth ova by fish, but it must be assumed that they can be carried. Helminth ova will tend to settle to the pond bottom and thus there may be a greater tendency for them to be ingested by bottom feeding fish. The helminth *Ascaris*, the roundworm, with the most persistant ova among helminths, is probably of little importance in excreta aquaculture reuse systems.

Certain parasitic helminths may be transmitted through recycling excreta in aquaculture systems since their life cycles include aquatic organisms, either fish, crabs, or aquatic macrophytes, as intermediate hosts. However, several helminths have a restricted geographical distribution and a particular infection would be of concern only in an area in which it is endemic. Perhaps the most important species is *Clonorchis sinensis*, the oriental liver fluke. It is associated with excreta fed fish ponds and is intensively transmitted where fish are eaten raw or partially cooked. Infection occurs mainly in China; Korea; Taiwan, China; and Vietnam and prevalence may reach 60% locally. The use of raw nightsoil in ponds in China was formerly responsible for much liver fluke disease. The eggs are fragile and die if stored for a few days, so that 7 days storage of nightsoil prior to pond enrichment would help to control it. However, since there are other important hosts such as cats and dogs, prior treatment of human excreta may only partially reduce transmission. Cooking fish well destroys the encysted larvae as it does for other helminths, but most fish preservation and pickling techniques have little effect. *Opistorchis viverrini* and *O. felineus*, cat liver flukes, have a similar epidemiological pattern to *Clonorchis*, but geographical distribution is much reduced to certain parts of Thailand and the USSR. Both liver flukes and schistosomes can be controlled further by keeping vegetation on pond dikes under control to discourage the snail intermediate hosts. The fish tape worm, *Diphyllobothrium latum*, has copepod and fish intermediate hosts but is not associated with excreta fed ponds; it is especially prevalent in lakes and rivers in temperate countries.

The cultivation of macrophytes for human and animal feed fertilized with excreta is common in Asia. The metacercariae of *Fasciolopsis buski*, the giant intestinal fluke, and *Fasciola hepatica*, the cattle or sheep liver fluke, attach to various aquatic macrophyte species which, if eaten raw or partially cooked, lead to fluke infections.

Enteric bacteria and viruses survive for considerably shorter periods in seawater than in freshwater, although the survival of protozoan and helminth ova in the two environments is similar. However, the latter tend to settle, and intermediate hosts of most helminth parasites would not occur in seawater. The coastal water environment may thus cause fewer public health problems in excreta reuse than inland waters. However, edible shellfish such as oysters and mussels concentrate enteric bacteria and viruses in their tissues, and outbreaks of polio, hepatitis A, and diarrheal disease have all been related to the consumption of shellfish from fecally polluted water. Furthermore, *Vibrio parahaemolyticus*, which causes acute gastroenteritis, has frequently been isolated from marine fish, shellfish, crabs and prawns.

The reuse of sewage containing toxic chemicals such as heavy metals and various organics may lead to their accumulation by cultivated organisms, which may thus constitute a threat to public health. However, levels of heavy metals and pesticides in fish grown in sewage stabilization ponds have generally been reported to be within acceptable concentration level limits.

Environment for Disease Vectors

The construction of pond systems for the reuse of excreta may provide an environment for the breeding of vectors, particularly mosquitoes that may transmit disease agents that themselves may not be found in excreta, such as malaria and filariasis.

There is little danger of mosquitoes breeding in ponds containing fish, since mosquito larvae are consumed by the fish. Tilapia species, which are highly suitable for excreta reuse systems, consume mosquito larvae. In ponds used to raise fish, open water areas should be maintained since certain types of mosquitoes breed in association with aquatic macrophytes.

In aquatic systems used specifically to raise macrophytes, mosquitoes may be controlled. In a water hyacinth aquaculture system, the mosquito fish *Gambusia affinis*, which is tolerant of low dissolved oxygen, can be stocked to consume mosquito larvae. In a duckweed aquaculture system, a more or less complete cover of weed on the surface would inhibit mosquito breeding since the larvae would be prevented from surfacing for air.

SAFEGUARDING HEALTH

It would be ideal to cultivate aquatic organisms with excreta, following its treatment to eliminate all pathogenic organisms. Unfortunately, prior waste treatment to achieve complete pathogen destruction would seldom be feasible from an economic or even technical point of view. However, certain procedures should be followed, according to the schema presented in Fig. 11, to reduce to an absolute minimum the chances of any public health hazards due to excreta reuse.

The destruction of excreted pathogens is principally achieved by a combination of time outside the human host and temperature. No excreted pathogens can survive a temperature of 65°C for more than a few minutes, except spore

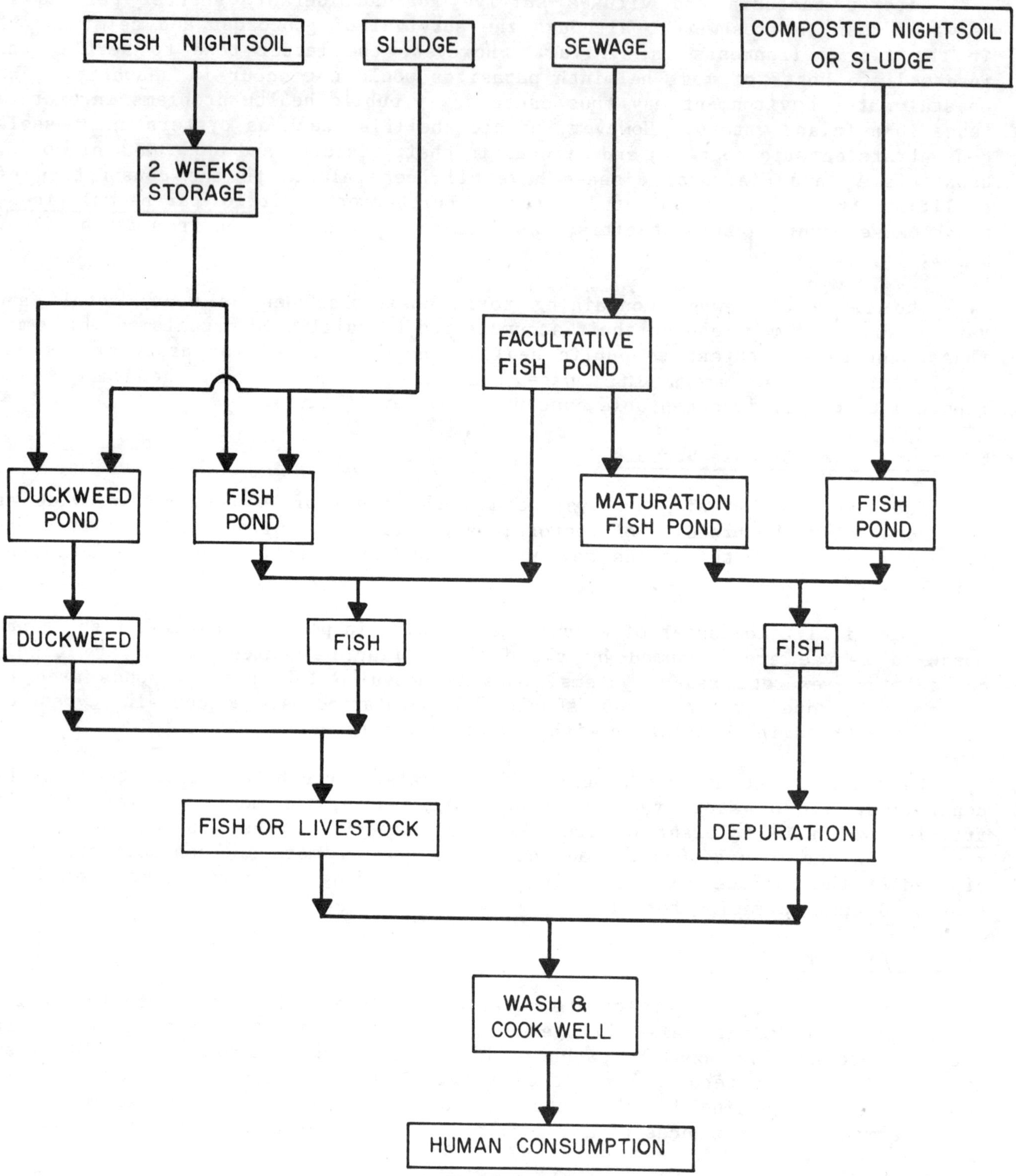

Fig. 11 Aquaculture reuse strategies with different types of excreta to safeguard public health.

forming bacteria and hepatitis A virus. Other extraintestinal environmental conditions that are important in pathogen attenuation are sunlight (UV radiation), high dissolved oxygen and high pH, but natural inactivation processes are very complex. It is disturbing that our knowledge of enteric virus inactivation is so limited, but attenuation is expected to occur rapidly in the tropics.

The only processes that produce a pathogen free material in treating nightsoil or sludge are batch aerobic or thermophilic composting, batch thermophilic digestion and drying for a minimum of two years. Only the first option is feasible as a low cost sanitation technology since the second is technically complicated and the third is too time consuming. A well designed aerobic composting system with good management is a viable, low cost sanitation technology.

Nightsoil, septage, or primary sewage sludge should never be added to a waste recycling system without storage for at least 2 weeks to eliminate any liver fluke or schistosome ova. Since cesspool and secondary sludges are at least partially digested, they may be recycled without further storage since any liver fluke and schistosome ova will have been destroyed. However, sludges, including effluents from biogas digesters, are still likely to contain significant densities of enteric bacteria and viruses.

Nightsoil and sludges should never be used in a waste recycling system to raise fish for direct human consumption; even though the amount of waste added to the fish pond may be relatively small compared to the degree of dilution by the water in the pond, fish tend to consume excreta directly and may thus ingest sufficiently high concentrations of bacterial and viral pathogens for them to enter the fish tissue.

Activated sludge and trickling filters are both poor technologies for pathogen removal from sewage, but a well designed system of stabilization ponds has a very high removal performance. A minimum of 3 ponds with a minimum total detention time of 20 days will produce an effluent that is either completely pathogen free or with only small concentrations of enteric bacteria and viruses; pathogen helminths and protozoa will have been completely destroyed. It is fortunate that stabilization ponds are both low cost and appropriate for warm countries, and are able to effectively treat excreta.

Fish should only be grown in waste stabilization ponds for direct human consumption if the concentration of bacteria in the water does not exceed a concentration of 10^4/100 ml. This is a much more realistic standard than that of 10^2/100 ml set up arbitrarily by WHO (1973) since the fish should be safe for marketing following depuration. Thus, fish essentially should be grown only in maturation ponds satisfying the above proposed standard for direct human consumption, and marketed following depuration. It should be emphasized that the standard proposed above, based on the findings of Buras et al. (1982), is only a proposal and should be tested further by the microbiological analysis of fish grown in maturation ponds. Although certain fish species, particularly tilapias, may grow well in facultative stabilization ponds, they should not be consumed directly by humans since they may contain pathogens in their muscle, organs, and intraperitoneal fluid; they should be processed and used for animal feed.

High rate stabilization ponds effect minimal pathogen removal since their detention time is so short, normally less than 7 days. At present, the only economically promising option is to feed the wet algal slurry to livestock, but this could lead to animal infection by the beef tapeworm, *Taenia saginata*, and *Salmonella* spp. for cattle; and *Salmonella* spp. for poultry.

Depuration should be incorporated with excreta reuse schemes in which fish are produced directly for human food, even in systems with adequate treatment, to eliminate the possibility of pathogens being present in the fish intestines or on the outside of the fish body. The best way to achieve this would be to harvest the fish and transfer them to clean water for a few weeks. An easier but perhaps less effective way would be to stop loading the system with treated wastewater or excreta for a few weeks. Fish which have been grown in water containing high concentrations of microorganisms may have pathogens present in muscle, organs, and intraperitoneal fluid, and depuration would then be ineffective.

It is essential to promote good hygiene in all stages of handling and processing excreta raised fish to minimize potential contamination. A final step which should render waste grown fish safe for human consumption is adequate cooking; the consumption of raw, undercooked or improperly processed or preserved fish should be discouraged.

It may be necessary to "lengthen the food chain" in most excreta reuse systems; i.e., organisms raised in excreta fed ponds should be harvested and used as animal feed (for fish, shrimp, or livestock) rather than being fed directly to humans. The addition of an extra step in the food chain eliminates the direct human consumption of organisms raised in excreta fed systems, and should lead to an additional safeguard to public health.

CHAPTER VI. ECONOMIC AND FINANCIAL ASPECTS

The emphasis in aquaculture is on the production of human food, but due to increasing energy costs and environmental pollution associated with conventional wastewater technology, there is considerable interest in the West in the potential of aquaculture as a simple and economically effective alternative for excreta treatment. It is becoming appreciated that aquaculture can provide an effective treatment alternative to conventional waste treatment and in addition provide an economic return to defray at least part of the operating costs. However, the economic potential for excreta reuse technologies has scarcely been explored, particularly for "dry" systems more relevant for developing countries.

ECONOMIC DATA OF DIFFERENT EXCRETA REUSE OPTIONS

Aquaculture is land intensive. Due to the increasing value of suburban land surrounding cities, efficient excreta reuse is not likely unless excreta can be transported for reuse into rural areas where land is relatively inexpensive.

For sewage, the tendency will be to treat the wastewater on as small an area of land as possible to minimize expenditure on land. The value of the products raised in a combined aquaculture sewage lagoon system is likely to be small compared to the overall expense of the wastewater treatment system. The Munich sewage fisheries were built on moorland of little agricultural value, and the Calcutta sewage fisheries, built on swamp, are rapidly shrinking due to city expansion.

Cartage sanitation options have a greater potential for excreta reuse since it is feasible to transport the stronger wastes to rural areas for recycling. Water transportation is the cheapest form of bulk transport if a network of canals or rivers is available.

There are relatively few data on excreta reuse. The composting toilet in Vietnam is reported to have a high degree of usage since the toilet is designed to produce compost with an economic value. The Indian National Environmental Engineering Institute calculated that the cash value of compost produced by a pour flush/double vault composting toilet was about US$10/yr, enough to repay the capital cost of toilet construction in 5 years.

Most of the existing economic data on excreta resource recovery through aquaculture is from Taiwan, China (Kuhlthau, 1979). In the city of Tainan, public and private nightsoil collectors there remove excreta from vaults and septic tanks, which are sold to fish farmers. Private collectors work only 10 months/yr, when there is a demand for excreta for fish ponds, but the public system works year round, and markets about 80% of the total collection. Illegal operators start nightsoil collection very early in the morning and use dippers and buckets for removal in conjunction with ox carts of 0.75 ton capacity. The public system uses 2-3 ton capacity vacuum trucks and laborers. During the peak season when nightsoil is at a premium, it is stolen from vaults before the municipal collection and thus there is in effect a black market in nightsoil. The public system sells the nightsoil for US$0.65/ton plus $0.57/km for transportation, while the nightsoil collected by private collectors is sold for $7/ton inclusive of transportation costs. The annual household excreta disposal costs in Tainan for cartage are as follows: household capital $9.6, household operation and maintenance $2.2, collection capital $2.2, collection operation and maintenance $15.3, a total of $28.3. Since the annual value of nightsoil for aquaculture reuse is $1.3, the net waste disposal cost is $27.0. Since the private operators must make a profit on their collection and sale of nightsoil, the public system must incur higher costs, or charge too little for excreta, or both. The trend of using excreta to fertilize fish ponds is decreasing because of the increasing availability of cheap commercial fertilizer and supplementary feed. The fish ponds around the city of Tainan are also being filled in by urban development, which gives a greater financial return on the land.

Nightsoil in Mainland China is reported to be expensive for use as a fish pond input and is mainly used to fertilize vegetables.

There appears to be a relationship between income levels and demand for nightsoil in Japan; Taiwan, China; and Korea, with per capita incomes of $3,600, $700, and $400, respectively. Very little nightsoil is still used in Japan, the most affluent of the three societies, perhaps 2% of all the nightsoil collected. In Taiwan there is still considerable demand for nightsoil but it is decreasing, while in Korea, the poorest of the three societies, nightsoil is still routinely used. Nightsoil is considered to follow the behavior of what economists call "inferior goods" since demand decreases with rising affluence. As farm incomes rise, there is an increased cost of farm labor and therefore more mechanization. It thus becomes more economical to use chemical fertilizers which are cheaper and easier to transport and utilize. Also, in Japan and Korea there are government subsidies for chemical fertilizers which lowers the demand for nightsoil further. However, in most societies in which excreta reuse could have a significant impact on sanitation, per capita incomes are low and substitute goods such as chemical fertilizers may be relatively expensive and in short supply.

ECONOMIC AND FINANCIAL COSTING OF SANITATION TECHNOLOGY

Once the sanitation technologies which are technically and socially acceptable to a given community have been selected, an objective comparative economic costing based on the actual physical conditions of the community should be made to aid planners and policy makers in the selection of the most appropriate technology. The technique of cost-benefit analysis is used to give the net economic effect of a given project, i.e., to develop a price tag for the project that represents the opportunity cost to the national economy.

The application of costing principles to sanitation projects is difficult since so little is known about the technology or costs of nonconventional sanitation options. A major problem with cost-benefit analysis, which makes it difficult to select the most appropriate sanitation technology, is that it is impossible to quantify most of the benefits of a sanitation program, e.g., improvements in public health and user convenience. However, excreta reuse may be quantified, and a sanitation option with a significant reuse component may be economically more feasible than one without resource recovery. An excreta reuse system would also create employment.

It is difficult to generalize in determining costs for a particular sanitation technology option due to considerable variations between countries for unskilled labor, land, water, and materials, the main inputs to sanitation systems. The single most useful figure for cost comparisons of technologies is the total annual cost/household (TACH), which includes both investment (mainly capital) and recurrent costs (mainly labor).

A distinction between investment and recurrent costs is important for both technical and financial reasons. A community with limited financial resources might find it impossible to raise the investment finance to construct a system with a large initial capital requirement, but could afford to build and maintain a system with the same TACH but with relatively high recurrent costs. Since most developing countries with poor sanitation have plenty of relatively cheap labor, cartage systems are the best solution from a financial point of view since they have a higher percentage of recurrent costs than sewerage. A further advantage with systems having high recurrent costs is that they have considerable scope for reducing costs in response to reduced demand since investment in new trucks can be delayed and fewer workers hired. Systems with high investment costs have little scope for reducing costs in response to reduced demand since construction of facilities is completed relatively quickly.

While economic costing of a particular sanitation option is of interest to planners, a consumer is more interested in financial costing, i.e., what he will be asked to pay for the system and how the payment will be spread out over time. Financial costs are subject to interest rates, loan maturities, central government subsidies, etc. The financial cost of a sanitation system to the consumer could be zero if the central government pays for it out of the general tax fund. For most onsite sanitation systems, ideally the consumer should pay for construction of the original facility, in a lump sum or through a loan at an interest rate which reflects the opportunity cost of the capital, and then pay a periodic sum to cover operation and maintenance expenses, if any; in this case financial and economic costs would be similar.

The government may be willing to subsidize part or all of the construction costs of a simple sanitation system to satisfy the basic sanitation needs of the people. Construction costs in the market place should be annuitized over the life of the facility at the prevailing market interest rate. If self-help labor can be used for part of the construction, the cost of hiring that labor should be subtracted from the total before annuitizing. Any operating and maintenance costs should be added to the total base financial cost and compared with household incomes to check affordability. If the technology is considered to be affordable by the target population, then financial arrangements can be made to help consumers to get loans from banks. If the technology's base financial cost is not affordable, then therewould be a need to compute an alternative set of financial costs which include a financial subsidy.

Pilot projects for various excreta reuse options should be set up to determine the economic value of resource recovery in each case. These would have a significant bearing on determining financial costing of sanitation options. Depending on the economic viability of a particular excreta reuse option, the consumer may be able to sell excreta to either the government or private sector for resource recovery. The sale of excreta could provide a financial incentive to use the toilet facility and could be used to defray construction and maintenance costs. To use a toilet when one has not been used previously, a person needs to perceive a tangible benefit. The financial return on sale of excreta or the products of the recycled excreta may indeed provide such an incentive to maintain the toilet system.

Municipalities normally lack the incentive and entrepreneurial skill to successfully manage a revenue generating operation. Since good management is needed to organize the collection, delivery and distribution of excretion in an aquaculture reuse system, and then to manage, harvest, and market the products, it may be wise for a municipality to contract out the waste reuse part of the system to the private sector.

It is often stated that a municipal reuse scheme should not be considered as a profit making operation, but merely a way to reduce costs and to motivate people to cooperate in sanitation schemes by demonstrating a tangible benefit. It thus follows that an economically appropriate test of a reuse system is not that it makes a profit, but only that its net cost is lower in terms of discounted cash flow than that of other sanitation options. If the private sector is to be involved, the municipality may have to pay the private firm a commission, based on the lowest competitive bid, rather than expect to sell a franchise.

CHAPTER VII. SOCIOLOGICAL ASPECTS

It is generally accepted that excreta reuse is desirable if public health can be safeguarded. However, there may be cultural barriers to be overcome since social values associated with sanitation may have a strong influence on the implementation of excreta reuse systems. Excreta reuse is not traditional in many countries since there may be deep rooted cultural prejudice against the consumption of fish reared on human wastes. The Chinese, Indians in West Bengal, and Indonesians, appear to have few objections to eating waste grown fish since they believe that they are safe for human consumption following gutting, washing and thorough cooking, but in general the Malay, Filipinos, and Thai would be reluctant to consume such fish. In Germany there do not appear to be problems in marketing fish from the Munich sewage fisheries.

Public health authorities in certain countries are vehemently opposed to the concept of excreta reuse because they are concerned that some pathogens may find their way to the handler and consumer. Such a negative attitude may actually impede developments in sanitation with associated improvements in public health, since the implementation of hygienic excreta reuse may provide a financial incentive to install or upgrade existing sanitation technology.

The challenge is to develop commercially viable excreta reuse systems in societies in which excreta reuse is not traditional, and to make it more widespread in those societies in which it already occurs. Certain aspects which may influence the acceptance of excreta reuse systems by societies and suggestions concerning pilot project implementation are discussed below.

SOCIAL ACCEPTANCE OF EXCRETA REUSE

Since excreta is regarded as a taboo in many societies, people whose occupation involves regular contact with excreta may themselves be avoided. Where cartage systems of excreta removal are used, as in much of the developing world, sweepers and nightsoil removers may belong to disadvantaged minority groups and live within segregated communities. In the Indian subcontinent, cartage is associated with the sweeper castes with untouchable status. However, a stigmatized occupation may be in demand where alternative sources of occupation are not available, or even more relevant for excreta reuse, when the occupation becomes more profitable by possible financial returns from excreta collection and reuse. Fishermen also have untouchable status in the Indian subcontinent, but members of higher castes became involved in aquaculture in Nepal because it was so profitable, and by doing so upgraded the status of fishermen. Low status is generally reinforced by low pay, so improvements in pay should lead to an improvement in status. A central tenet of excreta reuse is to establish a monetary value for excreta.

The offensive nature of cartage of nightsoil may be reduced by storage of excreta before handling, or by reducing the amount of handling, both of which should make the job more socially acceptable. Sludge produced by storage, and thus partial digestion, of fresh excreta or nightsoil is a liquid slurry with practically no odor. Perhaps a 2 stage vault with a preliminary settling chamber, in which primary digestion could take place, connected to a second chamber for excreta removal could be built cheaply. It should be feasible to design relatively simple appropriate technology to remove excreta from the toilet storage receptacle, to transport it to the reuse site, and to load it into the reuse system, with a minimum amount of handling. The use of improved technologies requiring less direct handling of feces should lead to an increase in the status of excreta removers and handlers.

In societies in which excreta reuse is not traditional and a resistance to the concept may occur, social acceptability of excreta reuse products may be achieved by lengthening the food chain, i.e., by using the excreta to raise fish or macrophytes which are not consumed directly by humans, but which are fed to livestock, fin fish, or shrimps.

IMPLEMENTATION OF EXCRETA REUSE

The difficulties in inducing changes in tradition bound societies are often overstated. The local people in a given community are often much more astute than they are given credit for; the main reason that they may not wish to implement a proposed new technology may be that they are not convinced that they will benefit from the change rather than that they have a general resistance to the implementation of new technology. Many peasant societies rapidly adopt new technology once they perceive that they will receive benefits, particularly financial ones, which are concrete and readily understood. The emphasis for the implementation of excreta resource recovery systems should be on economic returns and not on public health benefits which are more nebulous, even for the statistician, although health must be safeguarded.

In societies in which there is resistance to excreta reuse, there is generally a segment of society which is aware of the potential benefits of resource recovery and appreciate its value and necessity. Pilot project implementation should be carried out with the assistance of such local sympathizers. Once the economic benefits of excreta reuse have been demonstrated by a pilot project, it should be relatively easy to gain converts. The dissemination of a viable excreta reuse technology may be facilitated by bringing observers from other regions to view the system in actual operation. It is imperative that a pilot project site be carefully selected from a sociological as well as a technical point of view. An economically viable excreta reuse technology could fail if a pilot project is not implemented in a socially favorable site or if sociological aspects are not given due attention.

It is generally acknowledged that self-help schemes based on the spirit of self reliance of a community are best to achieve long term improvements in the community environment. Self-help schemes are based on the willingness of individuals to participate, but it may be difficult to get everyone to cooperate, particularly if positive benefits of the reuse scheme remain to be demonstrated. Such schemes are usually more effective in short rather than long term projects since it may be difficult to maintain enthusiasm unless there are significant benefits to the participants. The concept of self-help is probably best left to the dissemination of a given reuse technology, following the successful conclusion of a pilot project.

Improvements in sanitation may be encouraged by the enactment and enforcement of by-laws requiring households to provide themselves with latrines. Defecation in specific locations from which excreta may be collected is an obvious, necessary prerequisite to any excreta reuse system.

A careful market study should be conducted to see if the excreta reuse products are saleable. However, there should be few problems if the food chain is lengthened and excreta are used indirectly to produce food for human consumption. A more serious potential marketing problem (if the reuse system is not at the family or small rural community level) may be to guarantee that the products sold are always safe for human consumption. Subsistence farmers experience the whole agricultural cycle from production of the food to its consumption, and can thus supervise the management of the system. However, a commercial farmer who raises crops for a distant and impersonal market may adopt any practices, including shortcuts in the production cycle that may be a threat to public health, to increase his profits. In Thailand where it is unusual to see overhanging latrines on fish ponds because it is not socially acceptable to raise fish directly on excreta, a farmer when asked why he used such a toilet replied that his family did not consume the fish, which were sent to the market for sale. The urban consumer, who can only judge market produce by its appearance, must be protected from unscrupulous and unhygienic practices to ensure that the product is safe from a public health point of view. There must be enforcement of regulations to ensure that a given excreta reuse system is managed as designed to safeguard public health.

REFERENCES

Buras, N., B. Hepher, and E. Sandbank, 1982. Public Health Aspects of Fish Culture in Wastewater", Final Report, IDRC Sponsored Project.

Feachem, R.G., D.J. Bradley, H. Garelick, and D.D. Mara, 1983. Sanitation and Disease: Health Aspects of Excreta and Wastewater Management, World Bank Studies in Water Supply and Sanitation, Vol. 3, John Wiley and Sons, Chicester.

Feachem, R.G., and S. Cairncross, 1978. Small Excreta Disposal Systems, Ross Bulletin 8, The Ross Institute, London.

Kalbermatten, J.M., D.S. Julius, and C.G. Gunnerson, 1982. Appropriate Sanitation Alternatives: A Technical and Economic Appraisal, World Bank Studies in Water Supply and Sanitation, Vol. 3, Johns Hopkins University Press, Baltimore.

Kuhlthau, R.H., 1979. Country Studies in Appropriate Sanitation Alternatives, P.U. Report No. RES21, The World Bank, Washington, D.C.

McGarry, M.G., 1976. "The Taboo Resource, the Use of Human Excreta in Chinese Agriculture," Ecologist, Vol. 6, No. 4.

Pacey, A., 1978. Sanitation in Developing Countries, John Wiley and Sons, Chichester.

Shclef, G., and C.J. Soeder, 1980. Algae Biomass Production and Use, Elsevier/North Holland Biomedical Press, Amsterdam.

WHO, 1973. Reuse of Effluents: Methods of Wastewater Treatment and Health Safeguards, Wld Hlth Org. techn. Rep. Ser., No. 517.

World Bank Publications of Related Interest

Laboratory Evaluation of Hand-operated Pumps for Use in Developing Countries
Consumers' Association Testing and Research Laboratory

An invaluable guide for selecting water pumps. Paper rates 12 pumps on basis of rigorous testing to determine their strengths and weaknesses. Pumps rated for packaging, condition as received, installation, maintenance, repair, and overall desirability.

Technical Paper No. 6. 1983. 89 pages.

ISBN 0-8213-0133-0. Stock No. BK 0133. $3.

Laboratory Testing, Field Trials, and Technological Development

Contains results of laboratory tests carried out on twelve hand pumps for the World Bank and presents recommendations for improvements in performance, safety, and durability.

1982. 122 pages (including 3 appendixes).

Stock No. BK 9190. $5.

Meeting the Needs of the Poor for Water Supply and Waste Disposal
Fredrick L. Golladay

Examines economic, political, social, and cultural obstacles to meeting the needs of low-income people for water supply and waste disposal. Proposes steps to reduce these obstacles. Notes that communities organize and manage resources to carry out activities that cannot be profitably undertaken by households. Proposes methods to facilitate the formation and operation of community organizations to carry out these functions.

Technical Paper No. 9. 1983. 66 pages.

ISBN 0-8213-0238-8. Stock No. BK 0238. $3.

A Model for the Development of a Self-Help Water Supply Program
Colin Glennie

Presents one version of a practical model for developing, with high community participation, water supply programs in developing countries. Consideration is also given to sanitation program development and practical guidelines for program development are included. One of a series of informal Working Papers prepared by the Technology Advisory Group, established under UNDP's Global Project, executed by the World Bank.

Technical Paper No. 2. 1982. 45 pages (including 2 annexes, references).

ISBN 0-8213-0077-6. Stock No. BK 0077. $3.

Municipal Water Supply Project Analysis: Case Studies
Frank H. Lamson-Scribner, Jr., and John Huang, editors

Eight case studies and fourteen exercises dealing with the water and wastewater disposal sector.

World Bank (EDI), 1977. 529 pages.

Stock No. IB-0655. $8.50 paperback.

Notes on the Design and Operation of Waste Stabilization Ponds in Warm Climates of Developing Countries
J. P. Arthur

Practical design criteria for pond systems to fit various ambient temperature and reuse requirements, such as irrigation. Offers suggestions for start-up procedures, operation, and trouble shooting.

Technical Paper No. 7. 1983. 106 pages.

ISBN 0-8213-0137-3. Stock No. BK 0137. $5.

Ventilated Improved Pit Latrines: Recent Developments in Zimbabwe
D. D. Mara and Peter R. Morgan

Describes designs and materials for latrines, one for peri-urban areas costing about US$150, and a US$8 version for rural areas. Some 20,000 of the ventilated pit latrines have been built in Zimbabwe. They have been found very effective in eliminating odors and controlling fly breeding.

Technical Paper No. 3. 1983. 48 pages.

ISBN 0-8213-0078-4. Stock No. BK 0078. $3.

Village Water Supply

Describes technical aspects, costs, and institutional problems related to supplying water for domestic use in rural areas and proposes guidelines for future World Bank lending in this sector.

A World Bank Paper. 1976. 98 pages (including 4 annexes).

Stock Nos. BK 9099 (English), BK 9100 (French), BK 9044 (Spanish). $5.

Village Water Supply: Economics and Policy in the Developing World
Robert J. Saunders and Jeremy J. Warford

Addresses the problem of potable water supply and waste disposal in rural areas of developing countries where the majority of the poor tend to be found. Emphasizes the economic, social, financial, and administrative issues that characterize villae water supply and sanitation programs.

The Johns Hopkins University Press, 1976. 292 pages (including 4 appendixes, bibliography, index).

LC 76-11758. ISBN 0-8018-1876-1, Stock No. JH 1876, $22.50 hardcover.

French: L'alimentation en eau des communautes rurales: economie et politique generale dans le monde en developpement. Economica, 1978.

ISBN 2-7178-0022-0, Stock No. IB 0549, $7.

Spanish: Agua para zonas rurales y poblados: economia y politica en el mundo en desarrollo. Editorial Tecnos, 1977

ISBN 84-309-0708-4. Stock No. IB 0532. $7.

Water Supply and Sanitation Project Preparation Handbook

The three companion volumes guide planners, engineers, and community development and public health specialists working in projects related to water and sanitation in developing countries and help users prepare project reports. Case studies illustrate each phase of project development: identification, prefeasibility, and feasibility. Suggests approaches and methods of project evaluation to guide project team members.

Volume 1: **Guidelines**
Brian Grover

Technical Paper No. 12. 1983. 190 pages.

ISBN 0-8213-0230-2. Stock No. BK 0230. $5.

Volume 2: **Case Studies—Identification Report for Port City, Immediate Improvement Project for Port City, Prefeasibility Report for Port City**
Brian Grover, Nicholas Burnett, and Michael McGarry

Technical Paper No. 13. 1983. 352 pages.

ISBN 0-8213-0231-0. Stock No. BK 0231. $15.

Volume 3: **Case Study—Feasibility Report for Port City**
Brian Grover, Nicholas Burnett, and Michael McGarry

Technical Paper No. 14. 1983. 308 pages

ISBN 0-8213-0232-9. Stock No. BK 0232. $15. Stock No. BK 0367. 3-volume set: $25.

World Bank Studies in Water Supply and Sanitation

The United Nations has designated the 1980s as the International Drinking Water Supply and Sanitation Decade. Its goal is to provide two of the most fundamental human needs—safe water and sanitary disposal of human wastes—to all people. To help usher in this important period of international research and cooperation, the World Bank has published several volumes on appropriate technology for water supply and waste disposal systems in developing countries. Since the technology for supplying water is better understood, the emphasis in these volumes is on sanitation and waste reclamation technologies, their contributions to better health, and how they are affected by water service levels and the ability and willingness of communities to pay for the systems.

Number 1: **Appropriate Sanitation Alternatives: A Technical and Economic Appraisal**
John M. Kalbermatten, DeAnne S. Julius, and Charles G. Gunnerson

This volume summarizes the technical, economic, environmental, health, and sociocultural findings of the World Bank's research program on appropriate sanitation alternatives and discusses the aspects of program planning that are necessary to implement these findings. It is directed primarily toward planning officials and sector policy advisers for developing countries.

The Johns Hopkins University Press, 1982. 172 pages (including bibliography, index).

LC 80-8963. ISBN 0-8018-2578-4, Stock No. JH 2578, $12.95 paperback.

Number 2: **Appropriate Sanitation Alternatives: A Planning and Design Manual**
John M. Kalbermatten, DeAnne S. Julius, Charles G. Gunnerson, and D. Duncan Mara

This manual presents the latest field results of the research, summarizes selected portions of other publications on sanitation program planning, and describes the engineering details of alternative sanitation technologies and how they can be upgraded.

The Johns Hopkins University Press, 1982. 172 pages (including bibliography, index).

LC 80-8963. ISBN 0-8018-2584-9, Stock No. JH 2584, $15 paperback.

Number 3: **Sanitation and Disease: Health Aspects of Wastewater Management**
Richard G. Feachem, David J. Bradley, Hemda Garelick, and D. Duncan Mara

Summarizes knowledge about excreta, night soil, and sewage and their effects on health. Describes the environmental characteristics of specific excreted pathogens and the epidemiology and control of the infections these pathogens cause.

John Wiley and Sons (U.K.). 1983. 528 pages.

ISBN 0-471-90094-X. Stock No. IB 0558. $39.50 hardcover.

Appropriate Technology for Water Supply and Sanitation

The following volumes have been published in a less formal format. Some are abridgments of the three foregoing volumes.

Volume 1: **Technical and Economic Options**
John M. Kalbermatten, DeAnne S. Julius, and Charles G. Gunnerson

Reports technical, economic, health, and social findings of the research project on "appropriate technology" and discusses the program planning necessary to implement technologies available to provide socially and environmentally acceptable low-cost water supply and waste disposal.

1980. 122 pages (including bibliography), Stock No. BK 9186. $5.

Volume 1a: **A Summary of Technical and Economic Options**
John M. Kalbermatten, DeAnne S. Julius, and Charles G. Gunnerson

A summary of the final report on appropriate technology for water supply and waste disposal in developing countries, a World Bank research project undertaken by the Energy, Water, and Telecommunications Department in 1976-1978.

1980. 38 pages.

Stock No. BK 0049. $3.

Volume 1b: **Sanitation Alternative for Low-Income Communities—A Brief Introduction**
D. Duncan Mara

Describes in nontechnical language the various low-cost sanitation technologies that are currently available for low-income communities in developing countries and presents a general methodology for low-cost sanitation program planning.

1982. 48 pages.

Stock No. BK 9189. $3.

Volume 3: **Health Aspects of Excreta and Sullage Management—A State-of-the-Art Review**
Richard G. Feachem, David J. Bradley, Hemda Garelick, and D. Duncan Mara

Provides information on the ways in which particular excreta disposal and reuse technologies affect the survival and dissemination of pathogens. It is intended for planners, engineers, economists, and health workers.

1980. 303 pages (including 14 appendixes, references).

Stock No. BK 9187. $15.

Volume 10: **Night-soil Composting**
Hillel I. Shuval, Charles G. Gunnerson, and DeAnne S. Julius

Describes a safe, inexpensive treatment method for night-soil composting that is ideally suited for developing countries because of its simplicity in operation, limited need for mechanical equipment, low cost, and effectiveness in inactivating pathogens.

1981. 81 pages (including bibliography, 2 appendixes).

Stock No. BK 9188. $3.

World Bank Research in Water Supply and Sanitation—Summary of Selected Publications

A bibliography summarizing the papers in the Water Supply and Sanitation Series, as well as the World Bank studies in Water Supply and Sanitation published for the World Bank by The Johns Hopkins University Press.

1980.

Stock No. BK 9185. Free.

Prices subject to change without notice and may vary by country.

The World Bank
Publications Order Form

SEND TO: **YOUR LOCAL DISTRIBUTOR** OR TO **WORLD BANK PUBLICATIONS**
(See the other side of this form.)
P.O. BOX 37525
WASHINGTON, D.C. 20013 U.S.A.

Date ______________________

Name ______________________

Title ______________________

Firm ______________________

Address ______________________

City __________ State ______ Postal Code __________

Country __________ Telephone (______) __________

Purchaser Reference No. ______________________

Ship to: (Enter if different from purchaser)

Name ______________________

Title ______________________

Firm ______________________

Address ______________________

City __________ State ______ Postal Code __________

Country __________ Telephone (______) __________

Check your method of payment.
Enclosed is my ☐ Check ☐ International Money Order ☐ Unesco Coupons ☐ International Postal Coupon.
Make payable to World Bank Publications for U.S. dollars unless you are ordering from your local distributor.

Charge my ☐ VISA ☐ MasterCard ☐ American Express ☐ Choice. (Credit cards accepted only for orders addressed to World Bank Publications.)

______________________ Credit Card Account Number

______________________ Expiration Date

______________________ Signature

☐ Invoice me and please reference my Purchase Order No. __________.

Please ship me the items listed below.

Stock Number	Author/ Title	Customer Internal Routing Code	Quantity	Unit Price	Total Amount $

All prices subject to change. Prices may vary by country. Allow 6–8 weeks for delivery.

Subtotal Cost $______

Total copies ______ Air mail surcharge if desired ($2.00 each) $______

Postage and handling for more than two complimentary items ($2.00 each) $______

Total $______

Thank you for your order.

IBRD-0053

Distributors of World Bank Publications

ARGENTINA
Carlos Hirsch, SRL
Galeria Guemes
Florida 165, 4th Floor-Ofc. 453/465
1333 Buenos Aires

AUSTRALIA, PAPUA NEW GUINEA, FIJI, SOLOMON ISLANDS, AND VANUATU
Info-Line
Overseas Document Delivery
Box 506, GPO
Sydney, NSW 2001
Australia

BAHRAIN
MEMRB
P.O. Box 22103
Manama Town 317

BANGLADESH
Micro Industries Development Assistance Society
G.P.O. Box 800
Dhaka

BELGIUM
Publications des Nations Unies
Av. du Roi 202
1060 Brussels

BRAZIL
Publicacoes Tecnicas Internacionais Ltda.
Rua Peixoto Gomide, 209
01409 Sao Paulo, SP

CANADA
Le Diffuseur
C.P. 85, 1501 Ampere Street
Boucherville, Quebec J4B 5E6

CHILE
Editorial Renacimiento
Miraflores 354
Santiago

COSTA RICA
Libreria Trejos
Calle 11-13
Av. Fernandez Guell
San Jose

CYPRUS
MEMRB
P.O. Box 2098
Nicosia

DENMARK
SamfundsLitteratur
Rosenoerns Alle 11
DK-1970 Copenhagen V

EGYPT, ARAB REPUBLIC OF
Al Ahram
Galaa Street
Cairo

FINLAND
Akateeminen Kirjakauppa
P.O. Box 128
SF-00101
Helsinki 10

FRANCE
World Bank Publications
66 Avenue d'Iena
75116 Paris

GERMANY, FEDERAL REPUBLIC OF
UNO-Verlag
D-5300 Bonn 1
Simrockstrasse 23

GREECE
MEMRB
24, Ippodamou Street
Athens-11635

HONG KONG, MACAU
Asia 2000 Ltd.
6 Fl., 146 Prince Edward Road, W. Kowloon
Hong Kong

INDIA
UBS Publishers' Distributors Ltd.
Post Box 7015
New Delhi 110002

10 First Main Road
Gandhi Nagar
Bangalore 560009

Apeejay Chambers, P.O. Box 736
5 Wallace Street
Bombay 400001

8/1-B, Chowringhee Lane
Calcutta 700016

7/188, 1(CA), Swarup Nagar
Kampu 208002

Sivaganga Road
Nungambakkam
Madras 600034

INDONESIA
Pt. Indira Limited
Jl. Sam Ratulangi 37
Jakarta Pusat
PO Box 181

IRELAND
TDC Publishers
12 North Frederick Street
Dublin 1

ITALY
Licosa Commissionaria Sansoni SPA
Via Lamarmora 45
50121 Florence

JAPAN
Eastern Book Service (EBS)
37-3, Hongo 3-Chome, Bunkyo-ku 113
Tokyo

JORDAN
MEMRB
P.O. Box 3143
Jabal
Amman

KENYA
Africa Book Service (E.A.) Ltd.
P.O. Box 45245
Nairobi

KOREA, REPUBLIC OF
Pan Korea Book Corporation
P.O. Box 101, Kwangwhamun
Seoul

KUWAIT
MEMRB
P.O. Box 5465
Kuwait

MALAYSIA
University of Malaya Cooperative Bookshop, Limited
P.O. Box 1127, Jalan Pantai Baru
Kuala Lumpur

MEXICO
INFOTEC
San Lorenzo 153-11, Col. del Valle Deleg. Benito Juarez
03100 Mexico City

MOROCCO
MEMRB
2 Rue Moliere Racine
Casablanca

THE NETHERLANDS
Medical Books Europe, BV (MBE)
Noorderwal 38,
7241 BL Lochem

NIGERIA
University Press Limited
Three Crowns Building Jericho
Private Mail Bag 5095
Ibadan

NORWAY
Tanum Karl Johan, A.S.
P.O. Box 1177
Sentrum
Oslo 1

PAKISTAN
Mirza Book Agency
65, Shahrah-e-Quaid-e-Azam
P.O. Box No. 729
Lahore 3

PANAMA
Ediciones Libreria Cultural Panamena, S.A.
Av. 7, Espana 16
Panama Zone 1

PERU
Editorial Desarrollo SA
Apartado 3824
Lima

PHILIPPINES
National Book Store
701 Rizal Avenue
Metro Manila

PORTUGAL
Livraria Portugal
Rua Do Carmo 70-74
1200 Lisbon

SAUDI ARABIA
Jarir Book Store
P.O. Box 3196
Riyadh 11471

SINGAPORE, TAIWAN, BURMA
Information Publications Private, Ltd.
02-06 1st Fl., Pei-Fu Industrial Bldg.,
24 New Industrial Road
Singapore

SPAIN
Mundi-Prensa Libros, S.A.
Castello 37
28001 Madrid

SRI LANKA AND THE MALDIVES
Lake House Bookshop
P.O. Box 244
100, Sir Chittampalam A. Gardiner Mawatha
Colombo 2
Sri Lanka

SWEDEN
For single titles:
ABCE Fritzes Kungl. Hovbokhandel
Regeringsgatan 12, Box 16356
S-103 27 Stockholm

For subscription orders:
Wennergren-Williams AB
Box 30004
S-104 25 Stockholm

SWITZERLAND
Libraire Payot
6 Rue Grenus
Case postal 381
CH 1211 Geneva 11

TANZANIA
Oxford University Press
P.O. Box 5299
Dar es Salaam

THAILAND
Central Department Store
306 Silom Road
Bangkok

TRINIDAD & TOBAGO
Systematics Studies Unit
55 Eastern Main Road
Curepe
Trinidad, West Indies

TUNISIA
Societe Tunisienne de Diffusion
5 Avenue de Carthage
Tunis

TURKEY
Haset Kitapevi A.S.
469, Istiklal Caddesi
Beyoglu-Istanbul

UGANDA
Uganda Bookshop
Attn: Mr. Martin-Luther Galiwango
P.O. Box 7145
Kampala

UNITED ARAB EMIRATES
MEMRB
P.O. Box 6097
Sharjah

UNITED KINGDOM AND NORTHERN IRELAND
Microinfo Ltd.
P.O. Box 3
Alton, Hampshire GU 34 2PG
United Kingdom

VENEZUELA
Libreria del Este
Aptdo. 60.337
Caracas 1060-A

WESTERN SAMOA
Wesley Bookshop
Apia